Self-Sufficiency
Beekeeping

Self-Sufficiency

Beekeeping

Joanna Ryde

Skyhorse Publishing

Skyhorse Publishing books may be purchased in bulk at special discounts for sales promotion, corporate gifts, fund-raising, or educational purposes. Special editions can also be created to specifications. For details, contact the Special Sales Department, Skyhorse Publishing, 555 Eighth Avenue, Suite 903, New York, NY 10018 or info@skyhorsepublishing.com.

www.skyhorsepublishing.com

10 9 8 7 6 5 4 3 2 1

Library of Congress Cataloging-in-Publication Data

Ryde, Joanna.
Beekeeping : self-sufficiency / Joanna Ryde.
p. cm.
Includes index.
ISBN 978-1-60239-958-7 (hardcover : alk. paper)
1. Bee culture. 2. Bee culture--Equipment and supplies. 3. Honeybee. 4. Cookery (Honey) I. Title.
SF523.R93 2010
638'.1--dc22
2009046853

Printed in China

Disclaimer

The author and publishers have made every effort to ensure that all information given in this book is safe and accurate, but they cannot accept liability for any resulting injury or loss or damage to either property or person, whether direct or consequential or however arising.

CONTENTS

INTRODUCTION

So, you are thinking of keeping bees. This book will help you to enhance your self-sufficient lifestyle and make beekeeping a reality. Farming and eating your own honey is a pure delight; it is full of goodness and contains unique properties. If you are lucky, there will be lots of honey to sell at local outlets, or to give as gifts for family and friends.

Introducing bees into your life and officially becoming a beekeeper is an exhilarating and memorable experience: you are now the owner of a hive and responsible for 60,000 bees!

The new badge of Beekeeper brings, initially, a little apprehension. What have I done? Will the neighbors get stung? Will we ever be able to sit peacefully in the garden again? But don't worry, this rush of anxiety soon subsides and, with a few simple precautions to reduce risk, you will start to relax and enjoy the fascinating and absorbing life of the honeybee close up—and prepare for a taste of your very own honey!

Great comfort can be gained in having a mentor. If you are lucky and have a local beekeeper to befriend, not only will they be happy to pass on their knowledge, but perhaps some surplus equipment, too. If you are on a budget, then starting with good secondhand equipment will ensure welcome savings for you in those early days. It is a good idea to put off buying lots of new equipment until you are sure you want to continue with beekeeping.

Equally helpful and enjoyable will be going along to your local beekeepers' association meetings. Many associations also run informal evening classes where you can acquire basic knowledge about the honeybee, together with the skills to become a beekeeper. These classes are often run by local beekeepers, many with decades of experience, who are willing to pass on their knowledge to aspiring beekeepers.

More than anything else, beekeeping is about management, control, and learning to understand the honeybee. It can also become a very enjoyable and sociable pastime, visiting others' hives and picking up hints and tips. If you manage your hive well and the weather is favorable, you will, in a single season, be harvesting the most delicious honey you have ever tasted. It is a stunning—and encouraging—fact that a well-maintained hive with a strong colony can yield 40–80 lbs. of honey in a season.

In your new role as a beekeeper you will be spending time getting to know your bees, handling frames full of newly drawn honeycomb, and, if you are lucky, spying the queen bee among the thousands of workers and drones.

What could be more environmentally friendly than keeping bees? You will benefit and so will others as your bees busy themselves with pollination, helping farmers to achieve healthy crops, orchards to increase their yield, and local gardeners to grow a good quantity of summer fruits and vegetables.

Beekeeping encompasses so many interests, not least a fuller understanding of the countryside, and a better comprehension of wild and cultivated flowers and plants. You will discover new ideas on health, well-being, and food and, if inclined, there are a number of skills and craft techniques to learn, and to enhance your self-sufficient lifestyle.

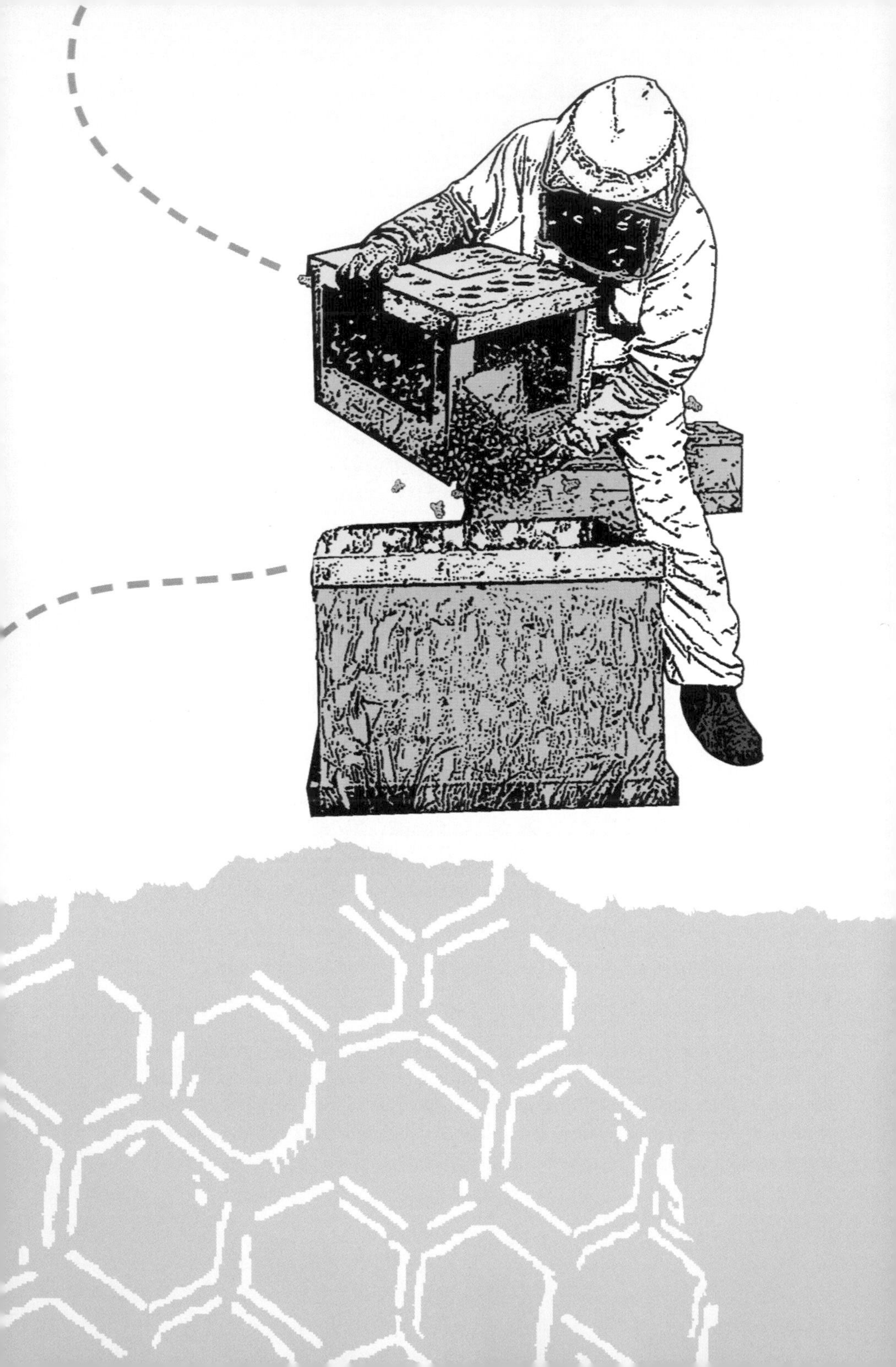

Keeping bees

Apiculture, or beekeeping, is the art of keeping bees with the aim of producing and harvesting the honey surplus and its other by-products. It is an ancient technique going back thousands of years.

Getting started

You may have been considering the idea of keeping bees for a long time, encouraged by childhood memories of watching beekeepers tending to their hives, or perhaps it is a recent interest encouraged by a local beekeeper with the tempting offer of a hive to get you started. Whatever your level of interest, it is important to do a little homework before bringing home your first hive.

The best way to learn about bees is to read as much as you can and to talk to experienced beekeepers on how they manage their hives. In the beginning, you will probably read and be given some conflicting advice that, when you are starting out with your bees, can be a little bewildering. Nothing beats the confidence gained from experience, however, and after your first year or two so much will suddenly fall into place.

Warning
It may sound obvious, but if you know you have an allergy to bee stings it is unwise to proceed. However careful you are there will be the odd time you get stung during a hive inspection. If any family member or friend suffers from anaphylaxis (an extreme allergic reaction), they should carry an epinephrine pen with them at all times while in the vicinity of an apiary.

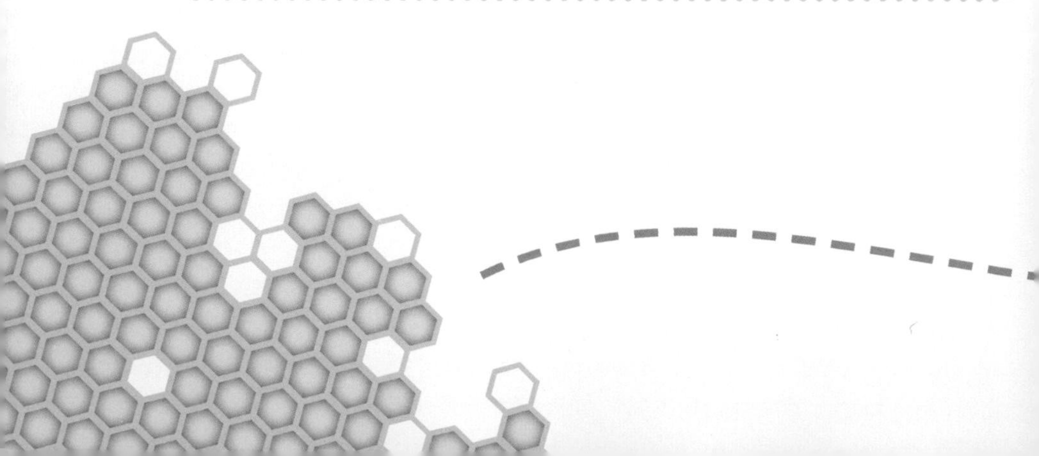

Join your local beekeepers' association

There are some things to consider and plan for that will ensure a smooth and encouraging start to beekeeping. Finding a mentor, especially in the early days of beekeeping, is invaluable. You will pick up lots of useful tips and have someone to call on in moments of confusion or doubt. An easy way to meet other like-minded people and to learn from experienced local beekeepers is to go along to your local beekeepers' association. Members come from all walks of life but have a common enthusiasm for beekeeping and a willingness to pass on their knowledge. Many are amateurs with one or two hives, but others will have many more colonies in their apiary. Regular meetings are held with expert speakers coming from around the country to give talks and demonstrations.

Some associations run courses during the winter months with weekly meetings for those interested in learning more about keeping bees. It is an invaluable and entertaining way to find out whether this unique pursuit is for you. Driven on by others' enthusiasm it can be a delightful way to learn about a new and fascinating subject. By spring, the meetings are held at local apiaries where the newly acquired theory is put into practice. This is an exciting moment: for many this will be the first time they have donned a bee suit and looked inside an open hive. For some, this first encounter, at such close quarters, can be overwhelming, but for most it soon subsides with the realization that they are well protected in their suit. At this point, the enthusiasm to get involved in the practical tasks rapidly takes over.

Get partnered with a mentor

With the theory and practical learning behind them, many budding beekeepers will go on that same spring to start with their first hive or two. Others will continue to complete a full season of beekeeping under guidance, either by continuing to visit the association's apiary or a fellow beekeeper's colonies, to help with the regular hive inspections.

If you choose to get on with setting up your own beehive, finding someone willing to guide you when it comes to some of the practical hive manipulations is a great way to learn and an enjoyable and sociable one too. Having a mentor to call in moments of uncertainty will make for a happier start to beekeeping.

The best time to start keeping bees

Spring is probably the easiest and most satisfying time to set up your hive. By starting at this time of year, you will give yourself time to settle into a routine with your bees, making your regular inspections with the bonus of watching the gradual build-up of comb and honey. If the next few months are managed well and you are lucky with the weather you will, by the end of the summer, be able to harvest your first crop of honey. If you complete a beekeeping course during the winter months, starting in spring is ideal if you want to keep up the momentum, enjoy the summer getting to know your bees, and be rewarded with the huge pleasure of eating your own honey—all within a few months of starting!

The bees, dormant through the winter months, become more active with the warmer spring weather. The colony starts to grow and the spring flowers, which start to bloom in March and continue through to May, together with trees, provide ample nectar and pollen. It is about this time that a prime swarm will take place (see Swarm control, p.56). If a swarm is discovered, a local beekeeper is often called to collect it. The caught swarm (a nucleus) is then available to be transferred to an unoccupied hive. If you make yourself known to local associations and beekeepers, it will probably not be too long before you are offered a nucleus of bees to start your hive.

In spring, bees are also readily available from beekeepers who are looking to reduce the number of hives in their apiary. If you are lucky you can be instantly set up by receiving a hive from a reputable beekeeper with an already established, disease-free colony of bees. For those who have not had much practice handling a busy hive and would be happier with a gentle introduction, starting with a nucleus (see Setting up your hive, p.39) of bees i preferable. This way you are really starting at the beginning. With fewer bees to manage on inspection, it will be easier to observe and understand how the hive works, watching the colony grow with the fresh new comb.

A place to keep your equipment

Where to keep your beekeeping equipment is something to be carefully considered. You may start with very little: your bee suit, a smoker, and a hive tool. Within your first season, however, you are likely to gather more bits and pieces than first anticipated. Bulkier equipment such as spare supers and frames, which you will need to add to your hive as the honey flow speeds up (see Setting up your hive, p.38) will take up space, and as your apiary and your experience grows you may want to acquire your own wax and honey extractors. As well as a place to store your equipment, you also need to think about where you can perform your honey and wax extraction (see Time to harvest, p.71).

Consider your neighbors

Most beginners do not have any choice in where they will keep their bees other than at home in the garden. If you have a larger than average garden you will probably have no problem placing your hive so that your neighbors never know it is there. With smaller gardens there is a little more to consider where neighbors are concerned. The last thing you want is for your bees to cause a nuisance and prevent the neighbors from enjoying their garden. Another potential annoyance is bees "spotting" clothing that is out on the line to dry, resulting in little yellow spots. In addition, if you or your neighbors have a pond, you will notice, some time in early spring, that the water's edge will be visited by large numbers of bees collecting water. They do this in order to break down and eat the old supplies of last year's honey stores. For some pond owners this can be fascinating, but for others it becomes troublesome. Put your neighbors' minds at rest by encouraging them to join you in one of your hive inspections. Get them involved by introducing them to some of the amazing facts about the honeybee and how vital their role is in the pollination of plants and crops. The promise of a pot of honey, or two, will certainly help.

Positioning your hive

A little planning before deciding where to place your hive will save time and needless stress later on. Keep in mind what matters foremost: the safety of your family and neighbors.

Ideally, a hive should be positioned facing southeast. This is to ensure that the rising sun beams onto the doorway of the hive encouraging the bees to make an early start to the day's foraging. Do not worry if this is not possible, however, as it will not make too much difference to your build-up of honey. It is worth noting though that to encourage a healthy colony, it is preferable to locate the bees in a sheltered area with a sunny spot away from persistent winds and drafts.

If there is no water nearby, ensure you keep a dish of water topped up by the hive. Including some stones and mossy sticks perched at the sides will help the bees to get to the water without falling in and drowning.

If you can nestle your hive between some bushes and tall hedges (not actually touching it—you want to keep the hive airy and dry with enough space for you to attend to it easily), away from the house it will attract less attention. The hive must be set on level ground which is easily achieved by bedding it in a flagstone. Adding some tall planting or screening ten feet or so in front of the hive is an excellent way to divert the bees' flight path upward, thus keeping them above both head height and neighboring gardens. It will also offer some welcome shade for the hives in the height of the summer. With all of these factors considered, you will be on the way to a successful season's beekeeping.

If you come to the conclusion that your garden is not suitable or big enough to keep your bees without it becoming troublesome, then there are alternatives to consider. Some people successfully, even in cities, keep their bees on the roof if they have an area that is stable and flat. Find out from other beekeepers where they keep their bees and you may find that they will be able to offer a space near their hives. Local associations often have an apiary and are in a good position to help you in locating a site.

Starting with one hive, or two

You could consider starting with two fully set-up hives or begin with one and have the other ready to take on a nucleus (swarm) of bees. By starting with just one hive, you are likely to feel a little more in control as you relax into managing the hive inspections and, equally important, you will have time to assess whether you are happy with the location you have chosen for your bees.

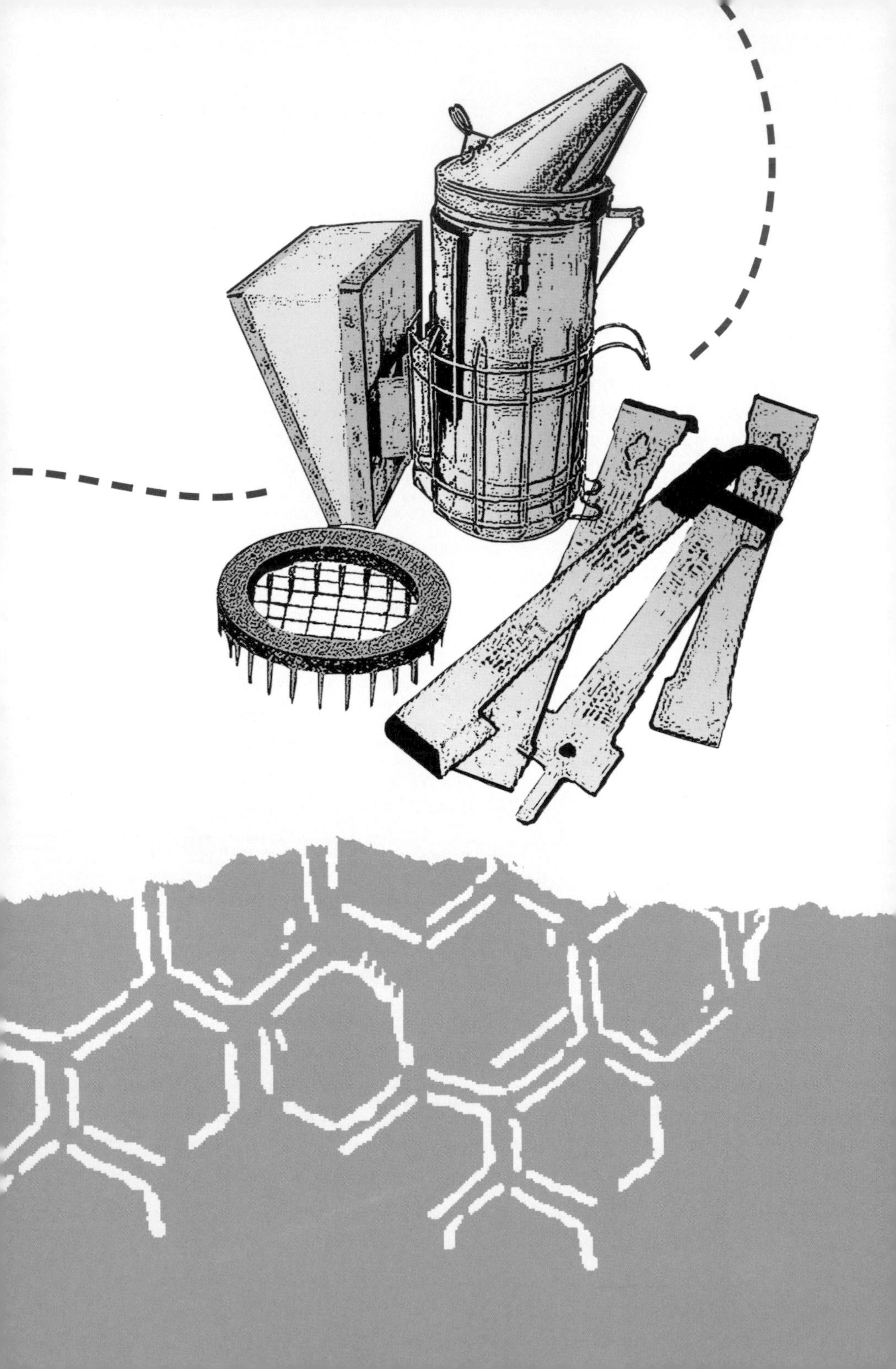

Basic tools and equipment

For the prospective beekeeper, flicking through suppliers' catalogs showcasing the diverse range of beekeeping equipment can be a little daunting. To the uninitiated, seeing so many unrecognizable tools and accessories can be overwhelming. Do not let this put you off because, as a beginner, all you need are a few key essentials; there is no need to invest in lots of complicated equipment.

Hive tools, equipment, and accessories

Hive Tool

The hive tool is something you will use every time you open the hive. It is indispensable and is used for levering open the sections of the hive, the frames, and excluder, which the bees will have sealed down with sticky propolis. You will have the hive tool in one hand throughout the inspection to loosen the frames before lifting them out, and then to dig out any undesirable features on the brood frames (see Hive maintenance, p.52).

The Smoker

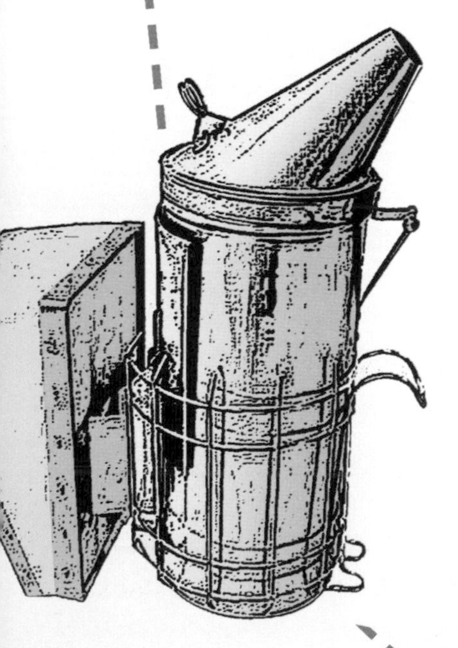

Recognizable to many, this is another must-have piece of equipment. A standard smoker will become your trusty friend, subduing a restless and agitated colony. A gentle puff across the bees will see them instantly clear the site you wish to inspect. To fuel a smoker, choose from old egg boxes, dry wood bark, corrugated cardboard, or sacking; any of these work well.

Frames

A ready supply of spare frames is essential when the honey flow is at its peak. Frames can be bought ready-made, but it is far cheaper and satisfying to buy them in kit form and simply put them together with a hammer and a few pins. If you do choose to attend an introductory course, you will be shown how to do this.

Beeswax Foundation

For every frame you use, you will need to fit a sheet of beeswax foundation. The foundation comes ready-embossed with the pattern of the comb impressed on the surface of the wax sheet as an encouragement to the bees to start creating new comb. Foundation is available wired and unwired. The diagonal wiring helps strengthen the foundation, preventing it from warping with the weight of either heavy brood or honey. If you are looking forward to cut-comb honey then be sure to use thin, unwired foundation. If you like your honey in a jar you will have to extract it from the comb, and for this you will need the added strength of wired foundation.

Spacers

These usually come as metal or plastic attachments that slip onto each end of the frame's top edge to determine the gaps between the frames and ensure the required space is allowed for a good build-up of brood and honeycomb. Landstroth frames are a very popular, and simple, type of self-spacing frame used in most beehives. An alternative used by some beekeepers are the Castellation Spacers. These come as galvanized steel or plastic strips placed along the inside of the super, into which the frames are placed at a set distance.

Feeder

At the end of the summer you will need to consider buying a feeder, which is filled with syrup and introduced to the hive during the autumn through to early spring, to replace the honey stores you have removed. The feeder will require occasional checking and topping off as necessary. There are a few types of feeders on the market, but rapid or contact feeders are common choices.

Dummy Boards

These are used for lessening a gap in the brood box when the frames are not spaced properly, acting as a false wall. The dummy board will prevent heat loss and build-up of brace comb.

Queen Press-in Cage

You may need to mark your queen, and to do this you should consider buying a queen press-in cage, which helps you to secure the queen bee while a small dab of color is gently placed on her (special marker pens are available). There are a couple of good reasons to "mark" your queen. It is so much easier to see her if she is given a small dot of color and it also denotes how old she is (see Seasonal reference guide: spring, p.112).

Porter Bee Escape

A porter bee escape is a useful little device for clearing the bees from the supers when you are ready to harvest the honey at the end of the season. It acts like a one-way door allowing the bees to travel down from the super you want to extract the honey from (see Time to harvest, p.70), while at the same time making it impossible for them to climb up and reenter. The porter bee escape works in conjunction with a clearance board, a sheet of plywood with raised edges, and a slot to hold the porter bee escape.

Straw Skep

To catch a swarm, you might consider buying a straw skep (hive). Before wooden frame hives were introduced, the straw skep was used to house the bees. Now they are used by beekeepers for catching swarms, and often you will see them on display promoting honey products. A sturdy cardboard box will work just as well as a skep for catching a swarm.

As your first season progresses, there will be other items for you to consider, some of which you may be able to borrow. Before you harvest your honey and wax you will need to prepare a supply of pots for the honeycomb and jars for loose honey. Also take time to consider whether you will need a comb cutter and or an extractor to release the runny honey (see Time to harvest, pp.71–73).

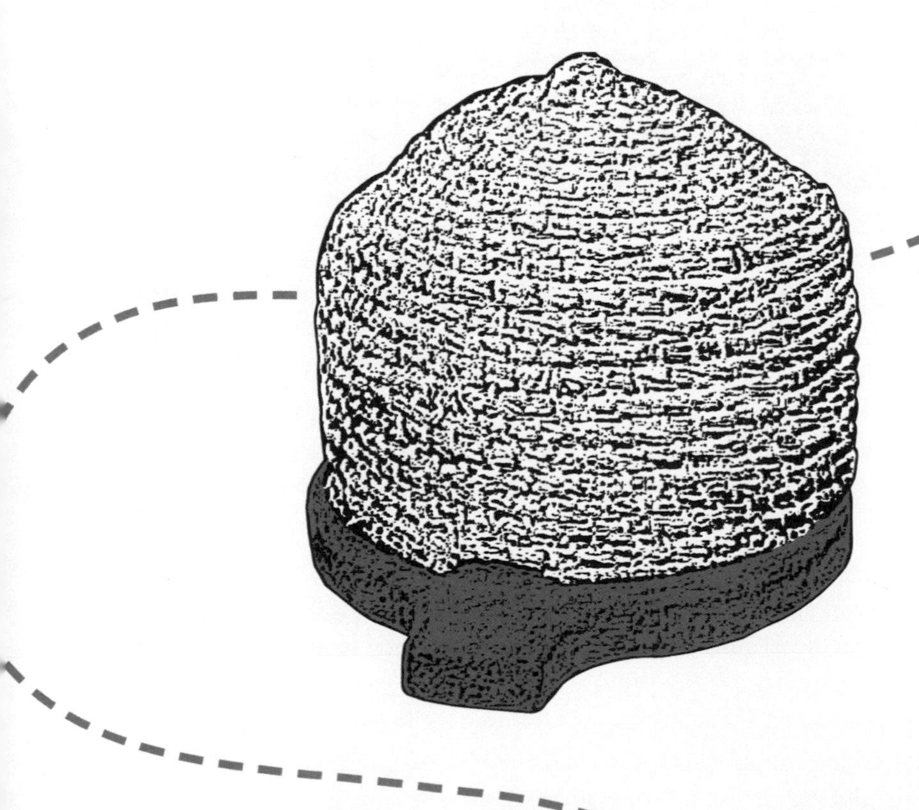

Personal equipment

Wearing a well-designed and correctly sized bee suit is vital for your safety. Knowing you are protected by the proper outfit will boost your confidence, and you and any visitor to your hive will feel more relaxed while viewing the bees.

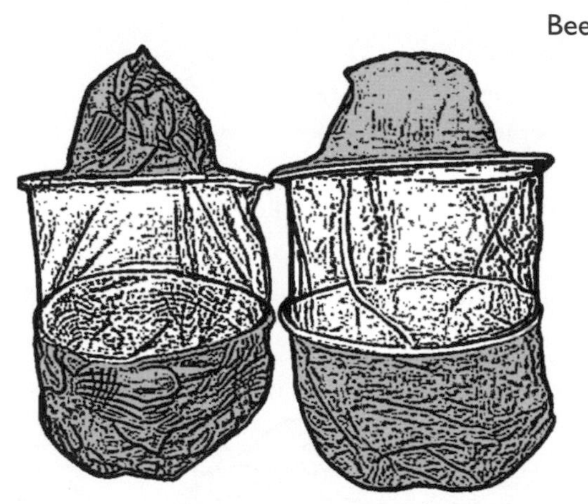

Bee suits are predominantly white and there is good reason for this. It is a fact that bees are less aggressive to pale colors than they are to darker shades. If you are bothered by a bee, it will be attracted to the darker features of your head, such as your hair, ears, eyes, nose, and mouth. You will notice when tending an open hive that the bees will be more attracted to the dark veil of your bee suit than any other part of it. The veil needs to be black, though, as it reduces glare, making it easier for you to see through.

The fact that bees are attracted to darker colors should be considered if you have any spectators at a distance who are not wearing bee suits. Never think that you can get away with putting on extra jumpers or thick woolly gloves instead of your bee suit as there is a chance you will be smothered during your hive inspection. If the bees become cross and agitated by an inspection, particularly when the honey is being harvested, it is because they assume they are under attack—and they will respond accordingly. Under these conditions, if woolly or furry clothing or objects (such as a child's toy, for example) are near the hive, they are likely to become a target.

There are various bee suits on the market to overcome any of these problems, from all-in-ones (full body suits) to separates comprised of a

Bear fact
Some beekeepers reason that there is an association for the bee with dark woolly items that goes a long way back to when bears which, of course, are dark and hairy, roamed the countryside and were a bee's number one enemy demolishing natural hives, dragging out the brood and honey comb to eat. Indeed beekeepers' hives in some parts of the world today where bears naturally roam, are constantly under attack from bears with a taste for honey.

jacket, trousers, and veil, and even camouflage outfits for those not wanting to attract attention while out on location. The all-in-one suit will cover you from head to ankle, providing effective and comfortable protection. They are easy to wear and work in, slipping over your regular clothing while remaining loose enough to avoid any precarious, tightly drawn areas! With an all-in-one, there is never any fear of the suit or veil inadvertently getting hitched-up, creating gaps that the bees could quickly find and enter.

The combined jacket, hat, and veil is another popular option. The jacket is attached to the veil and hat with drawstrings around the waist and wrists.

Gloves, gauntlets, and rubber boots are also essential to your safety. Gloves need to be either pale leather or of the plastochrome type. Your gloves will become very sticky with honey and propolis and will need to be washed or replaced regularly. It is surprising how easily a sting will penetrate thin leather, so do ensure your leather gloves are fairly thick. They may feel clumsier than a soft, pliable pair but they will offer much better protection. If you are unsure, wear a pair of latex gloves underneath for extra padding.

Gauntlets, whether they come attached to your gloves or as separate slip-overs, elasticated at both ends, will keep any gap covered, preventing bees from entering your sleeve or glove.

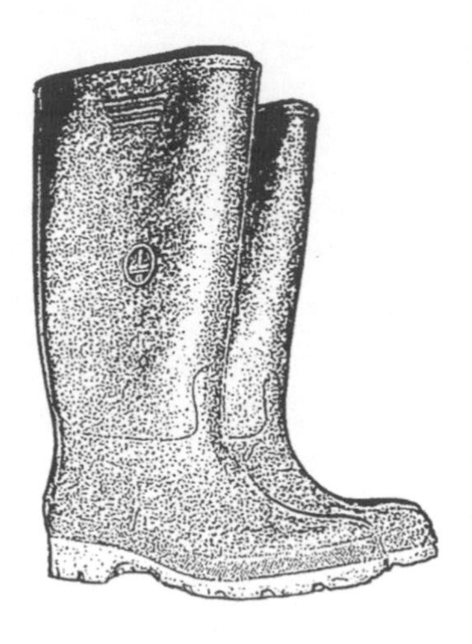

Even though most of your hive inspections will be done when the sun is at its highest and the day at its hottest, do not skimp on your protection. Rubber rain boots are not the most comfortable things to wear on a warm day, but they are probably the safest bet. With ankle boots there is always the possibility that the trouser leg will pull away from the boot, exposing your ankle, and the bees will sense this breach in your suit very quickly.

Veteran beekeepers are often pictured wearing just a hat and veil, handling the bees without gloves. A beginner should never even think about doing this.

If you do receive a sting, the speed with which the venom sac is removed, by scraping it away, will affect the severity of pain and swelling. It is also worth keeping a tube of antihistamine cream on hand in case you are stung.

Choosing your hive

Think of a beehive and you will probably picture the attractive WBC hive, which was at its most popular during the first half of the 20th century. Painted white, with a sloping roof and pyramid sections, it is a classic design, evoking strong childhood memories. It is tempting to emulate this romantic look, but do consider the options carefully before making that decision.

The consequences of choosing the wrong type of hive could result in your having to contend with heavy weights and particularly strong colonies; as a beginner this could become overwhelming.

There are a number of different hive designs to choose from that have been developed for different purposes. The most popular designs include the WBC, described above, the Modified National, the Langstroth, and the Commercial. Choose carefully as it could prove costly if you decide to start again with a new style of hive. Even within certain types of hives there is incompatibility with varying sizes of parts. Find out from local beekeepers which type they use and why. Apart from any other reason, it may be beneficial to use the same type as neighboring beekeepers for easy swaps and loans in the future.

The Langstroth

The Langstroth hive is probably the most widely used throughout the world except in the UK. In its "jumbo" format, which has extra space for deep brood frames, the Langstroth is a popular choice with commercial beekeepers. Be aware that it produces very strong colonies and can be heavy to handle when full of honey or brood.

Top Bar Hives

These are gaining in popularity with beekeepers who want to enjoy a cheaper and more bee-friendly method of beekeeping. Frames and foundation are not required and in turn, there is less heavy lifting to be done. Basically a series of bars are placed at the top of the hive and the bees do the rest. It is a very similar idea to that reported in the 1600s, describing the baskets the Greeks used, which simply had bars placed across the open top, covered with a lid. You will make more wax with this method but probably have a lower output of honey as it needs to be pressed; it is ideal for cut-comb, however. This method is widely used in developing countries and by experienced beekeepers interested in discovering new sustainable methods and perhaps experimenting for a more hygenic system of beekeeping.

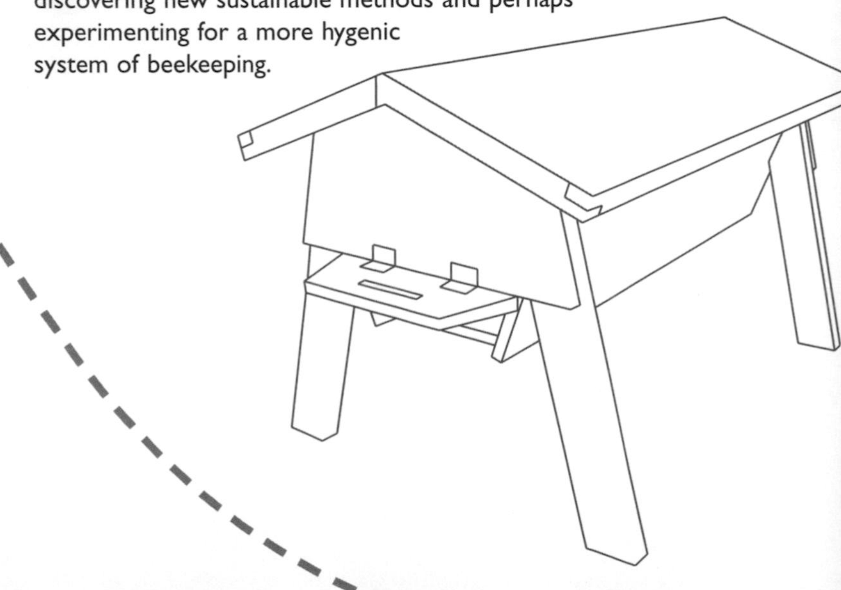

Modified National

The Modified National is possibly the most used hive throughout the UK and it is a good choice for the beginner. Its straightforward design makes it easy and not too heavy to handle when inspecting the hive. It is also less bulky than other hives, making it easier to store and transport. Its popularity means it is easier to come by good secondhand parts to build up your hive. When purchasing this type of hive, consider the open-mesh floor option as this will help to maintain important ventilation throughout the year as well as assisting with varroa mite inspections (see p.63).

The stand

A sturdy stand is an essential requirement of any hive. The hive needs to be raised off the ground to provide good ventilation, thereby lessening the risk of dampness creeping in. Do not risk using a stand if it is starting to show signs of rot or distorting out of shape as this could lead to the collapse of the hive when it becomes heavy with honey. It is also important that the hive sits in a level position. Placing your stand on a paving stone, that has been leveled into the ground works well. Inexpensive alternatives to the regular stands available, and commonly used by beekeepers, are timber beams, cinder blocks, and milk crates.

Bear in mind...

... that the height of your stand will determine the height of your hive, so consider how this will affect your hive inspection—will it cause backache by having to bend too much during inspections, or will you be over-stretching when it comes to lifting off the top super?

Building your own hive

If you have the time and skill, building your own hive is perfectly acceptable. Detailed plans are available from various sources including your country's beekeepers' association (see Resources, p.128).

Secondhand equipment

Buying, borrowing, or receiving secondhand equipment can help you make great savings with the initial set-up. You can track down secondhand equipment with the help of a range of sources, from regular and online auctions (see Resources, p.128), those retiring from beekeeping and dispensing with their equipment, or the offer of a fully set up, working hive from a beekeeper wishing to reduce his or her stock. If the offer of secondhand equipment comes along, knowing in advance what to look out for and asking the right questions could save you time and frustration later on. This is where having an experienced mentor or local beekeepers' association to call upon for advice is invaluable.

Whether or not you are on a tight budget, buying a new kit all at once can be a substantial outlay. Until you have really experienced handling bees it makes sense to avoid ordering lots of new equipment.

That said, care must be taken when buying secondhand equipment—hives in particular. If taking on a complete hive with frames, spacers, and queen excluder, it is not so much the shell that can cause problems as what is inside. Check for uniformity of frames and spacers that will provide the correct bee space (see Setting up your hive, p.38). Before introducing your bees to the hive make sure it is thoroughly cleaned to clear it of disease, mites, or wax moth larvae. A stiffrush to start with, followed by a gentle lick with a blow torch, will ensure nothing is left lurking in the corners or crevasses.

If a queen excluder (see Setting up your hive, p.36) is included in your purchase of a secondhand hive, examine it carefully as it needs to be in good condition. Over time, queen excluders can become bent and distorted. The perforations in an excluder are large enough to allow the worker bees, smaller than the queen bee, to travel up from the brood to carry out their work of building up the honey comb. If the queen manages to squeeze through due to a break or twist in the excluder she will slip up into your honey stores and spoil them by laying a brood. If you have any doubts, it is well worth buying a new queen excluder to avoid the disappointment of this happening.

Beginners' kits

If you really have no option but to strike out on your own and need to quickly get on with your set-up, then a beginner's kit might be the answer. Flat pack or ready-assembled hives are available and come with a range of standard kit included.

Mail-order services

As your skill and level of involvement grows, you will discover myriad utensils and gadgetry to help with every aspect of beekeeping. You will become aware of these as your requirements change.

Mail-order companies specializing in beekeeping equipment and accessories (see Resources, p.128) are readily available. Studying their catalogs and price lists is a good way to familiarize yourself with the various models of hive and tools offered.

Setting up your hive

The reason we keep bees in artificial wood-framed hives is that it makes it easier for us to optimize their environment and work toward producing a healthy and strong colony. The stronger the colony, the more worker bees will be produced to build the comb and to forage and this, in turn, will reward you with a good supply of honey and wax.

Happy bees

If bees are kept dry and draft free and have enough space for their brood nest and for the build-up of honey stores, they are happy to live in any suitable cavity such as a hollow in a tree.

When bees build a nest in the wild this natural hive comprises a number of spaced combs hanging down in parallel with the brood nest toward the middle and the surplus honey stores above. Honeybees have even been found happily building their hive in much more unusual environments, such as manholes in paved roads.

In the countryside, honeycomb has even been discovered in hedgerows, with the colony surviving into the winter, which shows that the honeybee may be far more resilient than we think.

Getting a new hive ready

If you could see a cutaway view of your hive you would see how it mimics the natural shape of the nest in the wild with the brood, pollen, and honey set in a specific pattern, albeit in a controlled environment.

Your wooden hive, with its detachable sections, echos the tiers found in a natural hive, but in a way that allows you to easily check the health of the brood, control swarms, and monitor your honey stores.

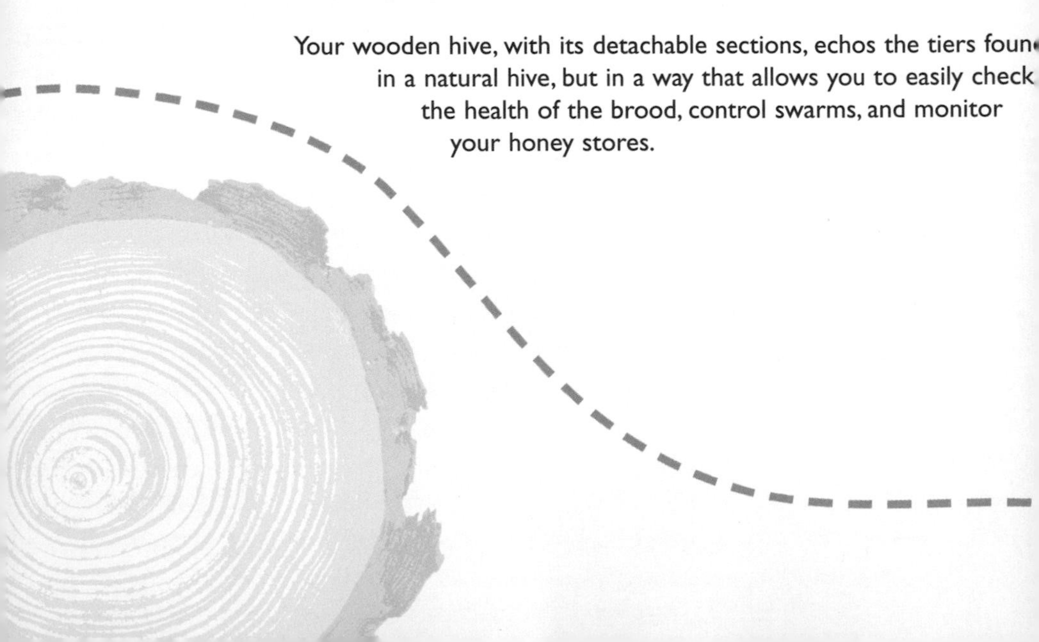

Working up from the bottom, a standard Modified National hive consists of:

1. At ground level a stand provides a firm base for the hive, elevating it to a comfortable height for inspections while also providing ventilation and security from unwanted intruders.

2. Next is the floor section, which incorporates the base of the hive, the entrance, and sometimes a landing board. In its most basic form, the floor is a solid timber base onto which can be added a tray and screen used to collect the hive debris, which you can pull out to make checks for varroa mites. A more recent innovation is the open-meshed floor. Unlike the traditional wooden floor, the base is made entirely of fine mesh that the bees can move around on while allowing improved ventilation and debris including mites to fall through to the ground below. Some of these open meshed floors also include a thin white panel which you can slide under the mesh to inspect the debris more closely for signs of disease.

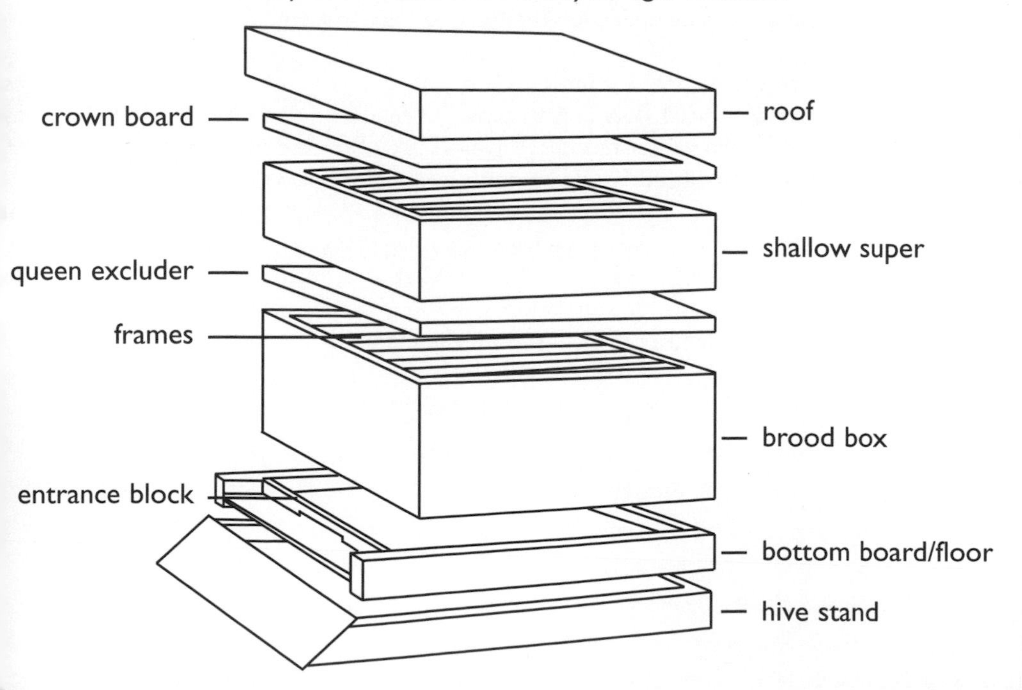

3. The brood box sits on the floor section, and this is the nest where the queen lives and lays eggs and the workers rear the brood. To allow for easy inspection and to give the bees a head start, the brood box contains brood frames. These are thin wooden frames that each hold a sheet of foundation wax upon which the bees will build the comb to hold the brood, some pollen, and honey.

4. A queen excluder is placed over the brood box to keep the queen from entering the honey stores. Without the queen excluder, the queen's natural instinct is to move up through the center of the hive to lay brood. The excluder is a screen with gaps that allow worker bees to pass but are too small for the queen to squeeze through.

5. The super, shallower than the brood box, is placed over the queen excluder. It is filled with frames, fitted with the wax foundation, which is where the bees will begin to draw the comb ready for filling with honey. As the season progresses, the honey flow becomes stronger and additional supers can be added for the bees to expand into.

6. The crown board is the covering placed on the top super. It has one or two elongated slots in it to allow for feeders to be placed over them during the winter months. The slots are also used for attaching the porter bee escape used for clearing the bees before harvesting.

7. The roof is the lid of the hive. The deep sides slip over the top of the super making it waterproof and acts as a safeguard against predators while maintaining gentle ventilation. The top is usually covered with a sheet of metal for extra protection from the elements.

Making frames and fitting the foundation

In preparation for your bees, the hive will need to be fitted with frames and foundation to encourage them to immediately start drawing new comb.

Frames can be bought ready-made or in kit form from your local or mail-order supplier. It will depend on the model of hive you have, but taking the National as an example, the brood box will usually accommodate eleven frames. The supers vary from eight to eleven frames. If you are using frames that have previously been used, take time to thoroughly sterilize them. Strip out the old, crusty wax, paying particular attention to the slots; these will help to hold your new foundation firmly in place while the bees build up the comb and fill it with nectar.

Putting new frames together from kit forms can be an enjoyable and satisfying task, but it is worth remembering that a single frame of honey can end up weighing 6–8 lbs. As a beginner, it is advisable to stop what you are doing occasionally to double check that you are putting the components of the frame together correctly before adding the foundation. Small pins (not glue) are used on the joints of the frame, creating a strong, rigid structure that will hold the foundation steady. If your frames are weak they risk collapsing into an irretrievable, sticky mess as the bees fill them with honey.

The foundation beeswax sheets usually come in packs of ten, cut to size and ready-rolled with an impression of the comb cells to stimulate the bees into action. As well as sheets of plain wax you can also purchase foundation that contains a single wire diagonally embedded in the wax. You should use this in the brood nest and in the supers if you are going to extract your honey. It helps to eliminate warping or collapse of the foundation in the hive and, if extracting honey, gives the comb additional support during the extraction process.

If you intend only to produce cut comb from your harvest of honey, a thinner, unwired foundation is available for this type of honey production.

Should you get to the point where you are retrieving a good quantity of wax from your hives then there is nothing to stop you making your own foundation (see Recipes and ideas, p.102).

Fitting the frames

Standing in front of your hive, you can either place the ready-to-use frames so that they are facing you head-on or fit them so they run from front to back. The usual way is to have them running from front to back, but whichever way you choose, it is important that the rest of the hive follows this pattern, running parallel up through the supers.

The "bee space"

When placing your brood and super frames into the hive, it is important that they sit parallel with the correct bee space. If the bee space between any two surfaces is too small for the bees to get through, the gap will be sealed with the gluey propolis that the bees produce. When the space is wider than a bee needs to crawl through, they will build brace comb bridging the frames together. This will make harvesting your honey very difficult without pulling apart and destroying the comb.

The bee is happy to travel through a space of 1/3 inch without manipulating it. If this is adhered to by using the correct spacers (see Basic tools and equipment, p.21) it will make brood inspections and comb removal clean, quick, and easy to complete.

Introducing a nucleus of bees

A nucleus of bees may be purchased from a reputable supplier or beekeeper. Although buying a nucleus can be an expensive way to start, it offers an easy and gentle introduction to beekeeping. The advantage over a swarm is that the hive will be smaller in numbers making inspections easy and enjoyable to carry out as you start. Your colony of bees will strengthen through the season, along with your confidence and skill in handling them.

A good nucleus, whether bought or inherited, should have at least three frames covered in bees with a good amount of healthy brood and stores, with the all important vigorous young queen.

On receipt of your nucleus, the bees should be hived as soon as possible, but before making the transfer, there are a few things for you to prepare ahead of time. Depending on how many frames your nucleus arrives on, the brood box will need to be fitted with enough frames to complete the "nest." To help the bees to progress quickly with the comb building, be ready with a feeder filled with syrup (see Seasonal reference guide, p.113). With your hive in position and ready for action, place the box holding the nucleus by the hive and allow the bees to fly from the box. This enables them to cool off and quench their thirst for a couple of hours before you prepare to transfer them early in the evening.

When the bees have resettled, have your smoker ready to give a few cool puffs across the top of the nucleus box to subdue the bees, then gently transfer the frames one by one to the brood box. Place them in the gap you have prepared in the center of the brood nest making sure they are set down in the same order as they arrived. Cover the brood box with a crown board and position the feeder accordingly. Put a super frame around this and cover with the roof.

The feeder will need a regular check every few days to see if the syrup needs topping off. After a week the bees will have settled and it will be time to make that first inspection and observe the exciting development of the brood nest. For the beginner, this is the perfect way to experience an open hive and to gain a good understanding of the bees inside.

Introducing a nucleus of bees into your hive by early May will stand you in good stead for achieving a strong colony by the end of the summer and, if lucky, a surplus crop of honey as your reward—all in your first season.

If you are unable to get started until later in the season, at the latest July, the bees will probably have just enough time to build up some stores for themselves to winter through, although they are likely to need more feeding through the winter than a more established hive. Timing is everything. If you want a strong colony and the thrill of harvesting your first honey by the end of the summer, you need to be sure that your nucleus of bees will arrive on time. A delay of just a few weeks can make all the difference to the rate at which your bees strengthen in numbers.

Starting with a swarm

If the cost of starting with a nucleus is something you would like to avoid, then taking on a swarm is another option, although it can present some difficulties to the inexperienced beekeeper. This is where having made contact in advance with local beekeepers and members from your local association is vital. The knowledge and assistance that an experienced beekeeper can bring is both helpful and reassuring, thus saving you a lot of time and worry. Word will travel that you are in need of swarm and it will not be long before you receive a phone call to collect your bundle of bees.

Swarms can be led by old queens who have been, or were about to be, succeeded by a new queen. It is not always easy to determine whether you are starting with an old queen who might not survive the winter, and dealing with this scenario could make for a trying start to beekeeping.

The steady hand of a mentor to advise and help you through the manipulations a swarm may require is invaluable. Trying to heed the advice of a number of different beekeepers with their own theories and practices can become a little confusing, especially when you are in the middle of a dilemma.

If you decide to proceed with a swarm, then the bigger it is—and they do vary in size—the better the odds of achieving a strong colony by the end of the season that will winter through well. Encourage the newly housed swarm by introducing a feeder to keep them nourished while they are busy drawing new comb to build the brood nest and foraging for vital stores.

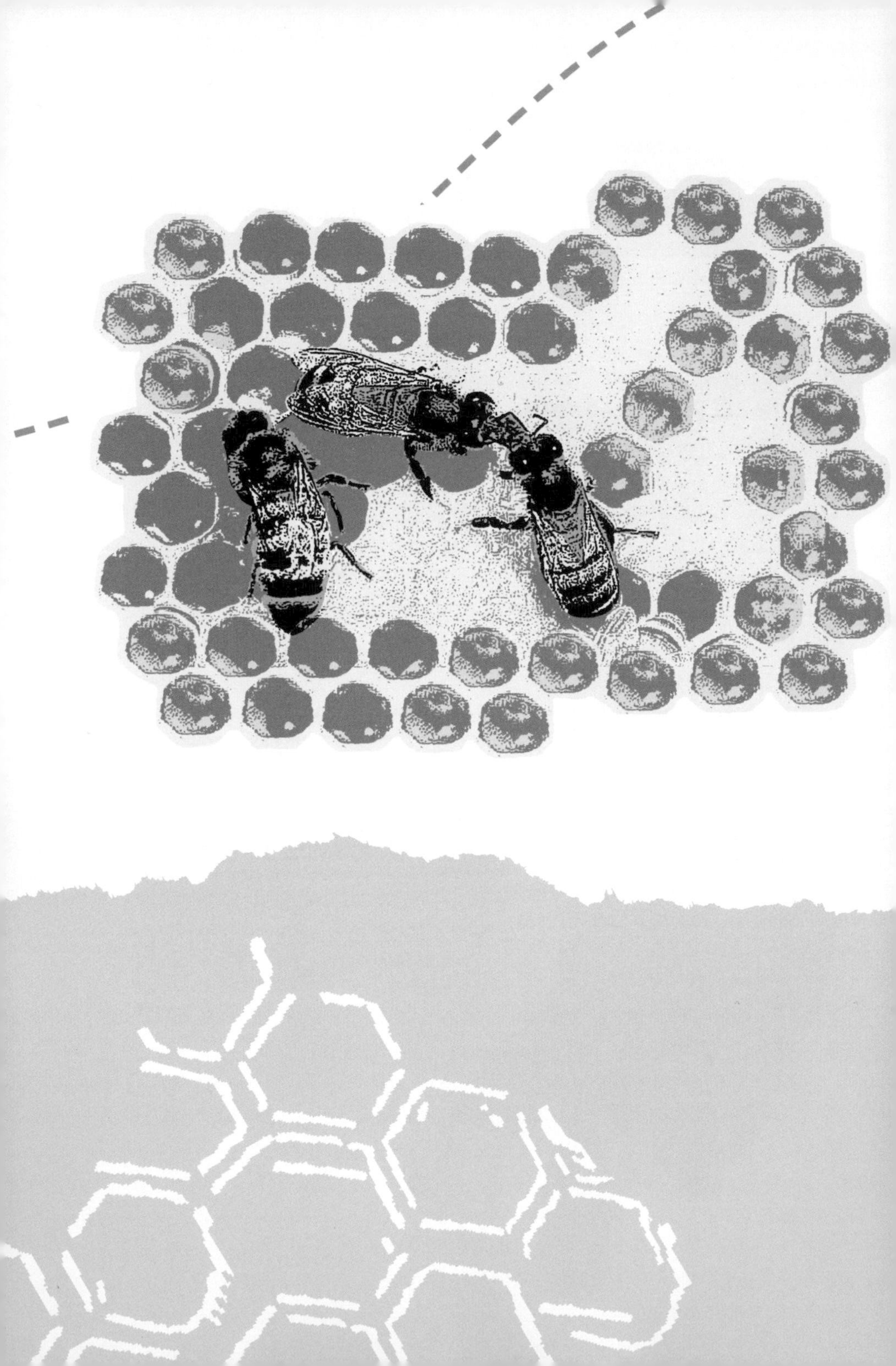

Inside the hive

Honeybees (*Apis mellifera*) are amazing insects and the more you learn about them the more bewildering the structure of a colony becomes. In one hive alone, a colony, at its summer peak, can have as many as 60,000 bees in it, so it is desirable to keep them in good humor. The workers make up the majority of the colony, along with a few hundred drones during the summer, headed by a single queen bee.

Hive hierarchy

An observation hive is an excellent way to appreciate the intricacies of a working hive. The unit is usually located inside a room with the bees completely on view and living happily cased between two glass panels. A tube running to the outside allows the bees to come and go. If you are lucky enough to come across one, they make for fascinating viewing.

Successful beekeeping is about good management, enthusiasm for the subject, and an understanding of the honeybee's life cycle. Learning to appreciate what makes for a happy and placid colony, and equally spotting the warning signs that will provoke your bees into abnormal behavior, will come with hands-on practice.

The queen bee

There is only one queen bee in a colony and it is she who determines the characteristics of all the bees in your hive, from their mood and to their healthy resistance to disease. She is larger than the other bees and has a distinctively longer abdomen than a worker or drone. She spends her life in the hive, apart from her brief departure before and during her mating flight and then again to lead a swarm to a new nest.

When the conditions are right, the young virgin queen leaves the hive for her mating flight. On this mission away from the hive she will mate with several drones that will supply her with sufficient sperm to lay eggs for the next three to five years. The queen is the only fertilized female and she is entirely responsible for laying thousands of eggs every day to ensure the survival of the colony. So important is the queen bee that she is closely guarded, chaperoned by her attendants, and fed on demand by the workers.

As a form of communication, the queen emits pheromones by which she alerts the colony that she is present; these chemicals influence the mood and behavior of the bees. Through the power of one ill-tempered queen a colony can become so aggressive that inspecting them could become unpleasant and tiresome.

When the queen is no longer in her prime, her egg-laying begins to slow down and when the colony senses the queen is failing they will take action to bring on a new vigorous queen. The appearance of two or three queen cells—sometimes up to twenty—is a sure sign that plans for supersedure are underway which, in turn, tells you that a swarm is imminent.

GRUMPY QUEENS

If you have a grumpy queen you will have an aggressive colony. Replacing your old queen with a young one from a docile strain will change the personality of your hive within hours.

The queen cells are much larger than the other brood cells and the nurse bees feed the larvae with considerable amounts of creamy royal jelly (see Time to harvest, p.84); this is the determining factor in turning the larvae into a queen and not a worker or drone. Instead of capping the cell in the manner of the rest of the brood, the queen cell is continually built upon until it is an elongated, tapered shape, with a pitted surface; it is often described as looking like the shell of a peanut. Within seven days the queen cell is sealed and capped with wax that the new queen will chew through when she is ready to emerge at around day sixteen or seventeen. When the old queen senses the emergence of a new queen she will swarm from the hive taking thousands of bees with her to form a new colony, leaving the original hive greatly reduced in numbers.

On the rare occasion that a queen is suddenly lost from the hive, which can happen if it is knocked over, the worker bees will take action, choosing a few regular cells that already hold young larvae to enlarge. These will then be fed large quantities of royal jelly, turning what was to be a worker bee into a queen bee.

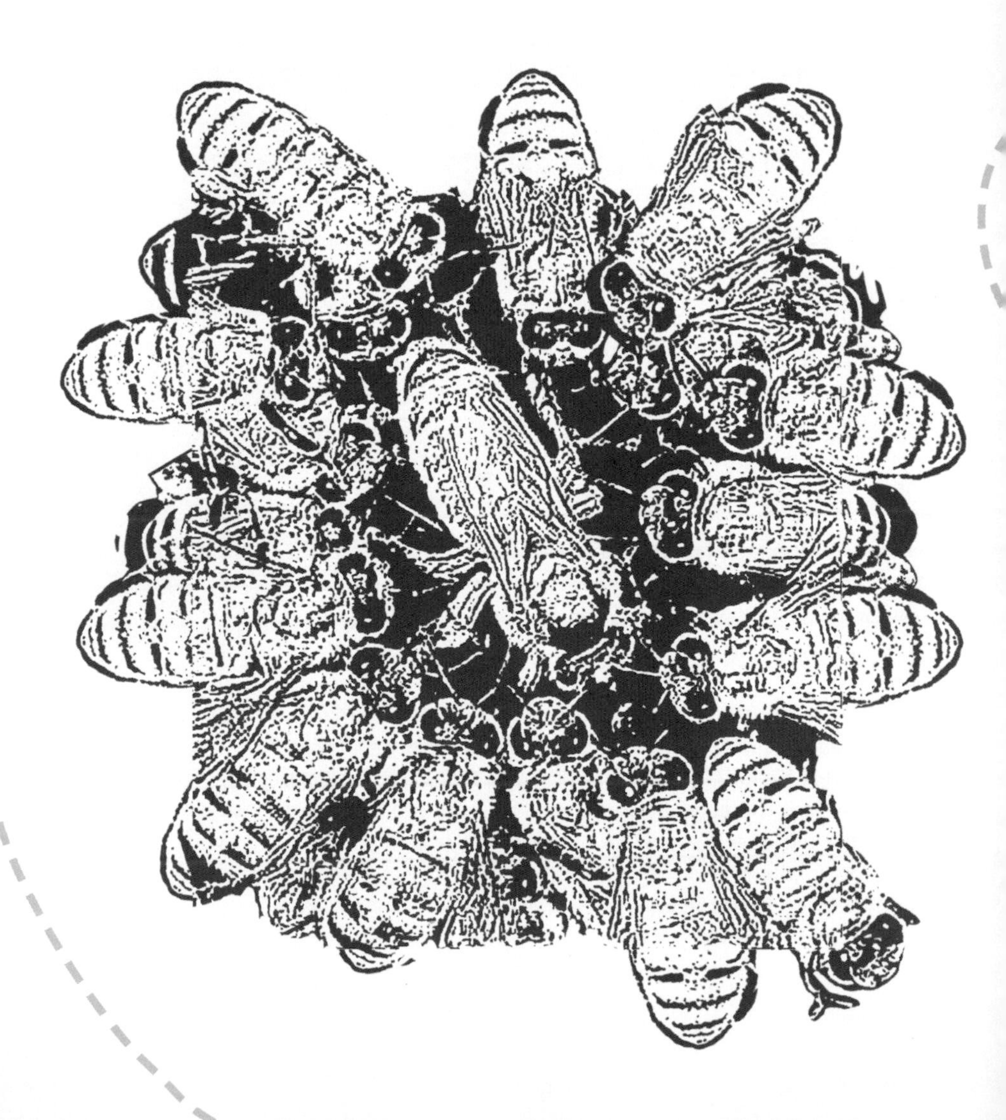

The worker bee

A worker bee is an unfertilized female. Despite being the smallest bee in the colony, she tirelessly carries out the greatest amount of duties in order to maintain the hive. Tasks are numerous and the type of job allocated correlates to the age of the worker.

During its first two or three weeks after emerging, the duties of a young worker bee are to clean the brood cells and clear any debris from the hive. They will also tend and feed the larvae and that of the developing queen, and when grubs turn into pupae, they will seal the cells. During this time, the young workers will also attend to the queen, escorting and feeding her. In addition, it is their duty to produce the wax that they need in order to draw new comb to receive and store the nectar that is brought back to the hive. When the comb is full of ripe honey the workers will seal the stores. Similarly, when pollen and propolis is brought back to the hive it is also the workers' job to pack away the pollen and use the propolis where and when it's required. By fanning at the doorway they also ventilate the hive and regulate the temperature inside. There will always be worker bees inside guarding the hive's entrance. When a worker bee is about eighteen days old its duties change and it will be required to scout and forage for nectar, pollen, and propolis, bringing back water, too. Before swarming, workers will have scouted for a suitable site on which to swarm.

It is clear that a worker bee performs furiously throughout its busy life. It might live, on average, for only five weeks during the honey flow, or longer if it survives to live through the winter. Its life span is largely determined by the weather and the state of its wings. Many simply wear them out through hard work, flying miles to forage for nectar and pollen and then unable to fly back to the hive.

DID YOU KNOW
that worker bees will fly as far as two or three miles to forage for nectar

When a worker bee finds an ample source of nectar it will fly back to the hive and communicate this discovery to the other foragers. By means of performing a "waggle dance," the bee can pass on the location of its find: inside the hive the bee will carry out the dance by waggling its abdomen and running in circles. From this display the other bees can determine where and how far away the nectar source is. Before flying out to collect the nectar, the bees feed on honey to sustain them on the journey.

When a flower has a good supply of nectar the worker bee will extract the watery solution by inserting its tongue into the flower for storing in

its honey stomach. Back at the hive it will regurgitate the thickening nectar, which the other workers will manipulate by adding an enzyme that they secrete to convert the sucrose into glucose and fructose (fruit sugar). This process thickens the honey, and excess water is also evaporated. The bees will help this process along by fanning air around the hive, creating quite a hum together with a delicious fragrant smell. Once the correct density has been achieved to avoid fermentation setting in, the workers will cap the honey cells, sealing it for storage (see Time to harvest, p.70).

Bees do not strictly hibernate during the winter; they remain in the hive clustered together, although moving around depending on the climate outside. To survive the winter the queen bee and the female workers live off the honey and pollen collected during the summer. For this reason, you should replace any honey stores you have taken with syrup, candy, or baker's fondant.

The drone bee

The drone, or male bee, is bigger than a worker bee, but not as big as the queen, with a heavier, rounder look, and no sting. Their only task in the hive is to mate with and fertilize young virgin queens. They have no other purpose in the colony and contribute nothing to the maintenance of the hive. They do not forage for nectar or pollen as they are fed by the worker bees and also help themselves to the honey stores. During the peak of the honey flow season a hive will hold several hundred drones but by the onset of autumn, when the colony prepares for winter, the worker bees will drive out the drones as part of their offensive to conserve honey supplies.

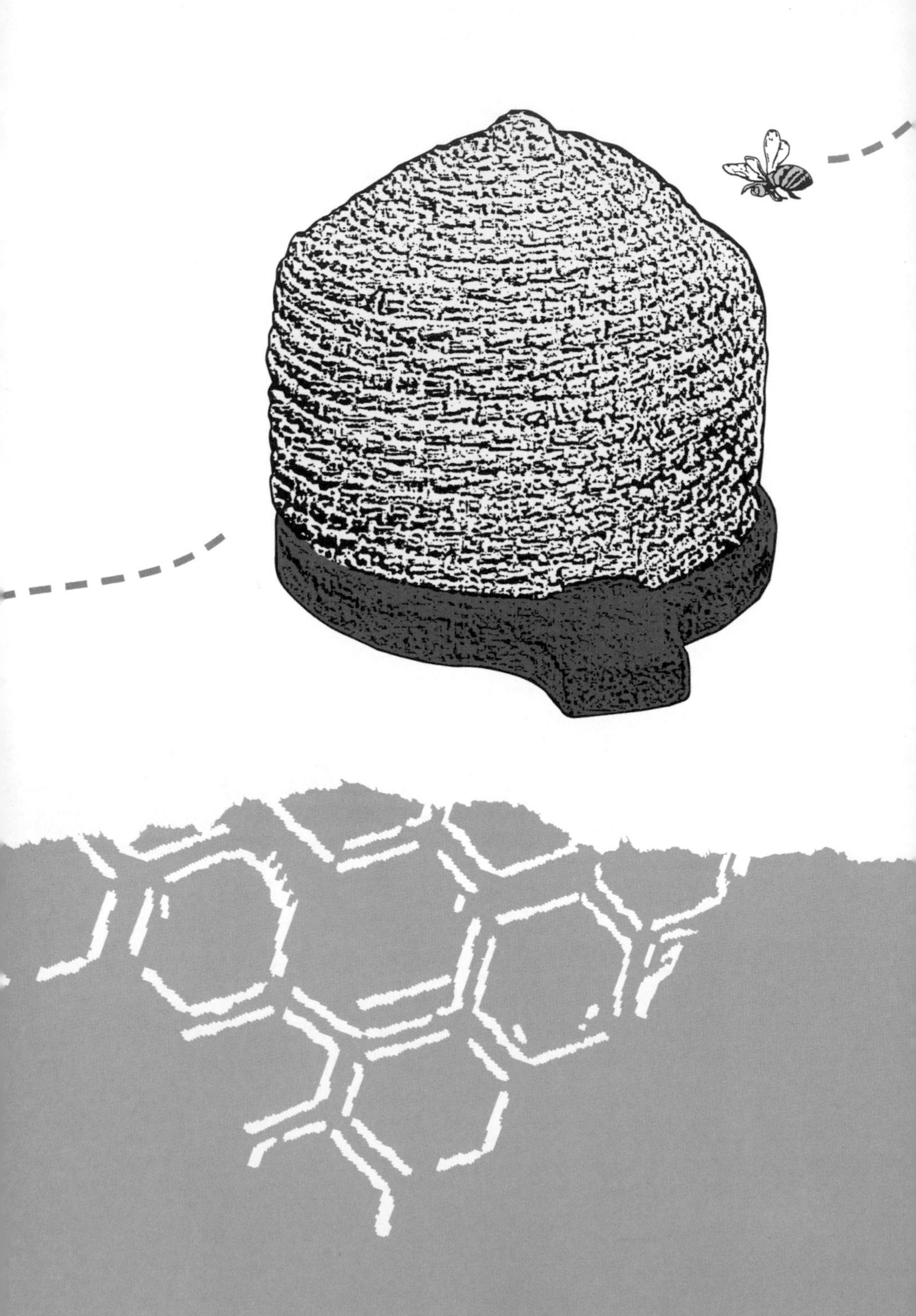

Hive maintenance

A well-maintained colony will help to provide you with a good honey crop. To ensure a strong, healthy colony, weekly inspections, particularly during spring and early summer, are essential. The swarming season is generally over by early July, and by August the honey flow is coming to an end.

Inspecting your hive

Every time you inspect your bees take time to prepare. Allowing yourself enough time is important because an unhurried and unflustered inspection will make a difference to the temperament of your hive. Check that your bee suit is fastened securely, however brief you think your visit will be. You must also ensure there are no gaps for the bees to crawl inside your veil, your boots, or your gloves. A few sensible precautions, which will become instinctive, will reduce the chance of being stung. A good smoker is an important tool; use it every time you tend to the hive. When a bee senses smoke, it instinctively plunges its head into the honey cells to gorge itself as it fears the hive is under threat. Light puffs of smoke will instantly clear bees and the control it brings will help your confidence. You will also need a hive tool to assist in levering and scraping off sections of the hive which will be sticky with propolis. It is also useful for pinching out unwanted cells from the frames.

Removing the roof and crown board

With the smoker lit and filled with enough fuel (cardboard or sacking) to last the session and a hive tool to help you break open the hive, you are ready to lift the roof off the hive and place it to one side. Underneath, covering the top super, is the crown board. This will be sealed to the super with propolis, the gluey substance that the bees secrete to close any gap they find in the hive. Gently ease the crown board with your hive tool. The gentler you are when handling the hive, the less agitated the bees will become. If you avoid abrupt movements and take care not to knock the hive, the bees should remain calm, making the rest of your inspection easier to complete.

Inspecting the supers

With the crown board removed, you will have exposed the top super (the sections where the bees build their honey stores) which will, depending on the time of the year, be a mix of new foundation, newly drawn comb, comb filled with honey but not capped, and honey ready to harvest. Always check there are enough stores for the bees particularly when checking early or late in the season (see Seasonal reference guide, pp.113 and 124). In the height of the season make sure there is enough room in the supers for the bees to keep busy building and filling new comb. A congested hive will be an unhappy and frustrated one and will be likely to swarm. If it looks like the bees will need more room, adding a new super will keep them content and busy (see Seasonal reference guide, p.118).

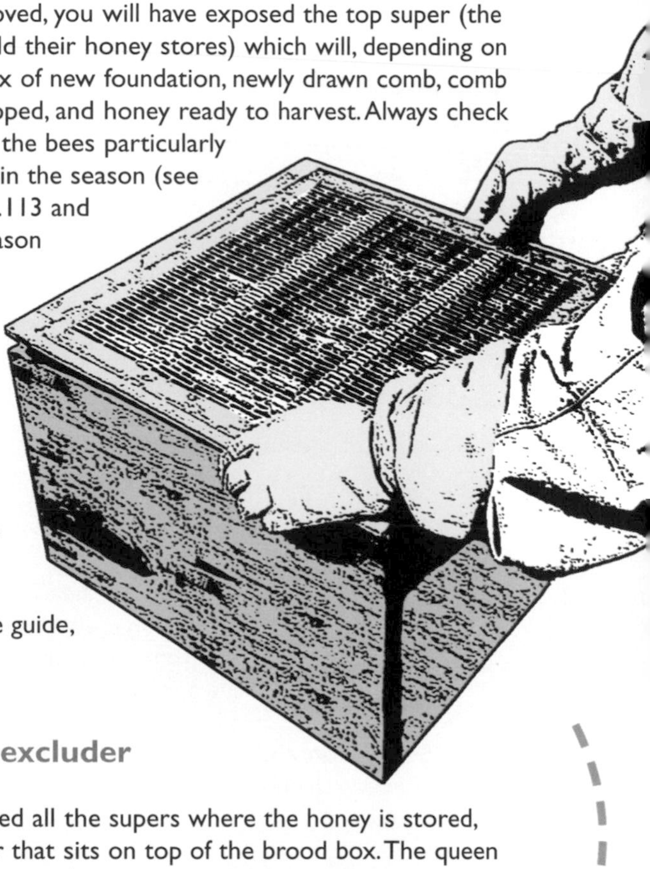

Importance of the queen excluder

By now you will have removed all the supers where the honey is stored, revealing the queen excluder that sits on top of the brood box. The queen excluder is a vital piece of equipment as this allows the workers to roam freely throughout the hive but keeps the queen below in the brood box. Without an excluder, the queen would naturally travel up through the hive laying brood in the supers, messing up your honey stocks.

Inspecting the brood

Your checklist should include:

1. Is there evidence that the queen is laying?

2. Search thoroughly for queen cups and cells.

3. Assess condition of brood and uniformity of brood pattern.

4. Look for signs of disease in the brood and any deformities in the bees.

You now need to inspect the brood, located at the base of the hive where the queen lives and lays her eggs. The brood nest is the most important part of any hive and each frame needs to be thoroughly checked. To inspect the brood, the excluder is carefully removed by gently releasing it from the brood box using the hive tool. The queen excluder will be sealed with propolis. Once loosened, carefully lift it off and check the underside for the queen before leaning it to one side of the hive. You do not want the queen to accidently drop out of the hive, as the bees will quickly swarm to find her. The likelihood of this happening is small, however, as the queen is shy and as soon as the brood box is exposed, she will run from the light and often a cluster of worker bees will gather round to hide and protect her. A gentle puff from the smoker at the door of the hive will push the bees up from the base; this will give you a better chance of glimpsing the queen as you work your way through checking the frames. Inspecting the brood is important for a number of reasons.

After checking the brood frames make sure they are replaced carefully in the position they were found. They will build a distinct pattern to the brood nest running throughout the frames and if changed, by shuffling them around, it will upset the bees.

Queen cells

Apart from checking the well-being of the brood and bees to ensure a healthy and growing colony, a critical reason for your weekly checks during the spring and summer months is to inspect the queen cups and find the queen cells (see Swarm control, p.58). If a healthy queen cell is left to mature, the old queen will decide to leave the hive, taking a swarm of bees with her, thereby radically reducing the numbers in your colony. This is not good news if you are hoping for a bountiful crop of honey and wax.

Drone comb

New beekeepers on their initial inspection of the hive are likely to confuse drone comb with a queen cell. Drone comb is often found at the base of the frame, or you may see the odd cell dotted throughout the frame of brood. Compared to the cell of a worker bee, the drone cell is larger, rounder, and more pronounced, but much smaller and blunter in shape than a queen cell.

Looking after your apiary

When your inspection is complete it is good practice to clear any old wax, pieces of comb, frames, and syrup from the site. Ensuring your apiary site is kept clean will help towards keeping down unwanted intruders, such as mice, wax moth, and woodpeckers.

Record keeping

Making use of a simple checklist each time you inspect a hive can be extremely useful and save you from needless confusion and frustration. By the time you have inspected your bees, put the hive back together, and cooled down from your inspection, trying to remember what you have noticed or what should be done on your next inspection will be impossible. The more hives you manage the more important the keeping of accurate records becomes.

Apart from the weekly observations, your checklist also needs to include longer term records of the age of the queen, how the colony performed in the previous season, and the general behavior of the bees. Keeping records simply makes the job of beekeeping easier and more enjoyable.

Here is a simple record sheet to give you an idea.

HIVE	INSPECTION DATE
Space	
Queen	
Brood/Eggs/Larvae	
Queen Cells	
Stores	
Disease	
Varroa mite count	
Action this time	
Jobs for next time	

Swarm control

Swarming is the honeybee's natural way of reproducing a colony and it is this reproductive instinct that has enabled the bee to survive from prehistoric times to this day. Each colony is driven to reproduce as many bees as it can and then divide and swarm to produce a new colony. For this reason, having complete control over swarming is difficult, but there are definite factors to watch for that will help to alleviate the problem. It is worth knowing that some strains of honeybee have a tendency to swarm more than others and so breeding from a queen from the appropriate strain is desirable. Talk to other beekeepers about their views and experiences with different strains of bees.

A major part, or indeed for many beekeepers the art, of apiculture is learning to understand how to prevent excessive swarming. Your aim as a beekeeper is to harvest a good amount of honey and wax and to do this, your hive ideally needs to maintain a strong colony throughout the summer to forage the high yields of nectar, to draw new comb, and to fill it with honey. Equally important, having strength in numbers will help a hive to winter through more successfully.

There are a number of causes that influence a tendency for a hive to swarm and several ways to deal with it. By the middle of May, sometimes as early as April depending on conditions and where you are located, when the honey flow is fairly substantial, the colony will be priming itself for a swarm. Queen cells will be formed and it is from one of these that a new virgin queen will hatch. If this is allowed to happen, the old queen will leave the hive with about half of the bees to find a new home. They will fill themselves with honey from the stores before leaving the hive, furiously flying around, usually at about midday or in the early afternoon. Within minutes the bees will descend and settle on a branch of a tree or bush, or in fact anything that takes their fancy, forming a large cluster that can vary in size from a tennis ball to a large soccer ball with as many as 20,000 bees or more, with the queen safely buried in the middle. After an hour or two, sometimes longer, the swarm will depart to its new home.

If you are not around when the swarm takes off you will be oblivious to what has happened. Only when you come to do your hive check might you notice that there are fewer bees, together with the tell-tale sign of a hatched queen cell or one that is about to hatch.

Keeping only young vigorous queens that have performed for only one or two seasons will keep the colony happy. They will be less likely to swarm than if headed by an older queen in her third season, whose activity in the hive will be on the decline.

The brood nest needs to be inspected every nine days in order to catch a queen cell before it hatches. By performing these regular checks you will be able to take preventative action against a swarm. Each frame must be carefully checked—top, bottom, and sides—for queen cells. They are easily missed if your concentration is broken or you accidentally skip checking a frame. Sometimes the bees will cluster to hide the cell, so gently finger them away to check what is underneath. The building of queen cups and queen cells in the brood chamber will give you a clue that the colony is restless and preparing to swarm.

On finding queen cups and queen cells, you will have to consider whether to remove the queen cells or instigate other initiatives to control swarming. There are a few things that can be done to lessen the bees tendency to swarm.

Queen cups are not necessarily a bad sign but check to see if they hold eggs or larvae, as this will be a queen in the making. Finding sealed queen cells will indicate that a swarm is about to occur or that it has already departed.

The only way to find out whether or not the colony has divided and swarmed is to make a search for your old queen. Having previously marked your queen with a dot of color will make the job of spotting her much easier, especially for the beginner. If you spot a queen with this marking you will be in no doubt that it is your old queen, as a young queen that has superseded her will have no marking (see Seasonal reference guide, p.117).

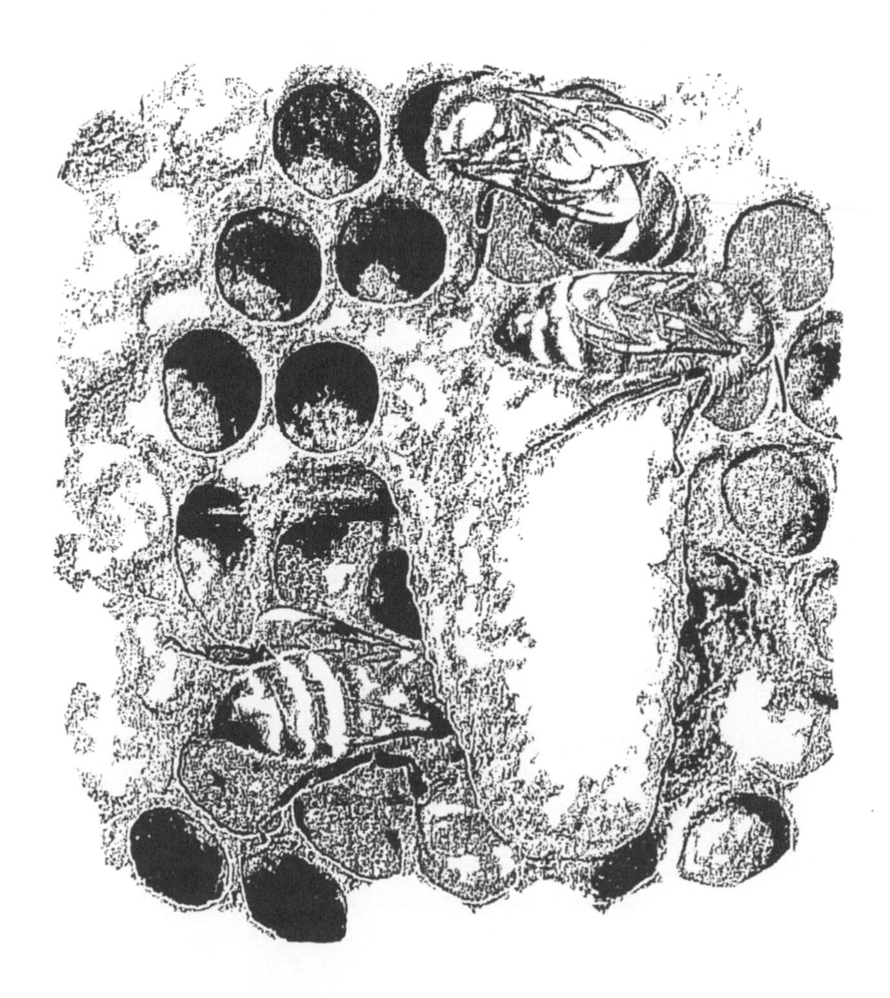

If you cannot find a queen and there are sealed queen cells in the brood nest, take a look at the condition of the brood. If you can see queen and worker cells in their early stages with eggs and larvae, continue to carefully pinch out all the sealed queen cells leaving one open cell with a healthy larva (mark this frame or, better still, like many a beekeeper, stick a drawing pin in the top of the frame as an indicator when you next inspect). The workers will create a new queen either by using this queen cell, or by adapting a worker cell that has an egg or young larva already in it. You are now back in control to continue with your weekly inspection.

Leave the original brood box for a week before inspecting it for brood and queen cells. At this stage you are looking for the healthiest of the queen cells—large, dimpled, and without any sign of damage. Having chosen your future queen you must, unfortunately, destroy the other queen cells.

Now leave your hive undisturbed for a couple of weeks, which will allow the young virgin queen to perform her mating flight (see Inside the hive, p.44). If an inspection coincided with the new queen performing her mating flight it could throw her into confusion and upset all your hard work, so it is worth the wait for that next inspection. Within three weeks you should see signs that your new queen has started to lay eggs.

An artificial swarm

If the old queen is still present in the hive and you want to avoid an imminent loss of a large swarm, then performing an artificial swarm (splitting the colony) will save your bees. You will need another hive for this manipulation. Lift out the comb the queen is settled on and place it in the brood box of the new hive. Place another two combs in the new hive with the bees on it and some stores and brood from the parent hive. Then remove the parent hive to a position six or more feet away. The new hive can then take up the position of the old hive. Once this is done, a large number of bees, but not all, will fly from the parent colony to join their queen in the new hive, satisfying their urge to swarm. Replace the missing frames in the parent hive with new ones with either drawn comb or new foundation. If you do not do this, and large gaps are left, the bees will build brace comb locking the frames together.

A congested hive

Bees need a job to do and a happy hive is a busy hive, so space is of paramount importance. Apart from foraging far and wide for their nectar and pollen, bees like to be kept active inside the hive with space to draw new comb in readiness to store the honey, and to busily nurture lots of healthy new brood.

A congested hive will cause problems so check that yours is not suffering from overcrowding. Adding a super as early as April will ensure the bees have room to start storing honey if the opportunity is there; otherwise the brood nest, the only other place to store supplies, will become congested with honey, taking away precious space that is needed for the brood. Since the brood nest is contained in the lower section of the hive, the bees will be agitated because they have no means of expanding the nest and this will stimulate them into a swarm mode. Replacing a couple of the old frames in the brood nest will also help alleviate the problem.

Overheating in the hive

A hive that becomes overheated during a spell of particularly hot weather can agitate the bees and become a contributory factor towards swarming. If it is possible to position your hive so that it receives a little shade in the middle of the day this may help, along with ensuring sufficient ventilation by having the entrance fully opened.

Maintaining ventilation throughout the whole year is important. As part of the honey-making process, bees need to evaporate water from the honey before capping the comb. Evaporation is also a way of keeping the colony at a constant temperature. Ventilation is especially important in the winter as a cold, damp hive is detrimental to the health of a colony.

Disease Control

Like any other creature, the honeybee is susceptible to a number of diseases. Knowing how to detect the signs of these infections in their initial stages will go a long way to helping you achieve an effective remedy. Taking preventative measures to avoid disease should become part of your regular beekeeping management. Always seek the advice of an experienced beekeeper or bee diseases officer if in doubt.

Colony Collapse

A relatively high-profile disorder in the United States is the cause of huge losses that have been reported in recent years. It is not unusual to experience colony loss during the winter but the magnitude of loss suffered by some beekeepers, and the consequences of a dwindling in bee numbers, is pushing researchers into action to discover the exact cause.

Wax Moth

The wax moth is found in hives from time to time. It is the larvae of the wax moth that cause damage by consuming large amounts of wax, destroying the comb and brood. Stored comb is susceptible to wax moth so it is advisable to check it before placing it in the hive.

Chalkbrood

Chalkbrood is a fungus that infects the brood; when consumed by the bee larvae it will germinate and eventually kill the bee. Diagnosis is easy—the contaminated cell becomes chalky white and mummified. It then loosens in the comb, allowing the worker bees to remove it. On most occasions it is not serious, but if it becomes so, re-queening could help to overcome it.

Varroa

Varroa is a common brood disorder. Visible to the naked eye, this crab-like mite can infest badly-managed colonies by feeding on the developing bee. Every colony will need treatment—commonly used methods are Apistan, Bayvarol, or Apiguard to keep it in check, together with integrated pest management as it becomes an increasingly virilent disease. Fitting a special varroa screen to your hive will reveal the level of infestation. Varroa favors drone the brood, so one way of checking to see if your bees are under attack is to uncap sections of brood cells: it will be apparent if you can see the reddish-brown or tan-colored mite.

American foul brood

This is a serious and destructive disease, which again affects the brood. American foul brood is caused by very resistant spores that attack the larvae, leaving them to die in their sealed cell and, as a result, leaving the brood combs themselves contaminated. The disease is easily spread by swarms from infected hives, infected honey, combs, and equipment. Control of this disease is through compulsory destruction of the infected hive.

European foul brood

This occurs when bacteria multiplies in the gut of an infected larva, starving it of nourishment. If early diagnosis is missed, the larva will contaminate the comb itself, allowing the disease to become rampant, debilitating the colony, and eventually wiping it out. This disease must be reported to your local bee inspector, and if badly affected, the brood must be destroyed. In some cases the bees can be treated but only by an appointed officer.

Dysentery

Inspection of a hive in spring may reveal signs of dysentery, presenting itself as brown streaks on the outside of the hive and on the frames inside. If bees are unable to make their cleansing flights due to long spells of cold and are exacerbated by feeding on indigestible stores, they have no choice but to excrete in the hive. When this occurs on a large scale the colony may die.

Acarine

This is caused by a tiny mite that invades the breathing tubes of young adult bees, causing death. There have been serious outbreaks in the past, but it affects relatively few colonies. Microscopic examination is the only way to diagnose this condition. Some strains of bee are hardier at resisting an attack from the mite, but the best way to avoid the condition is to keep strong stocks and maintain your regular checks.

Nosema

Ingested spores germinate in infected adult bees, producing millions of spores that are released through the feces causing further contamination. There are no obvious signs, so keeping strong, regularly inspected stocks is the best means of prevention.

The following checklist will help you to keep your hives healthy and avoid them succumbing to pests and diseases:

1. Read and learn about disease and pest management. Contact the American Beekeeping Federation for the latest news (see Resources, p.128). Advice on how to submit diseased bees for sampling is also available.

2. Learn how to make a diagnosis by distinguishing between a healthy brood and the abnormalities of an unhealthy brood.

3. Make hygiene a priority by keeping the apiary clear of discarded combs, lumps of wax, and propolis. Ensure secondhand equipment is thoroughly cleansed, scorching it with a blowtorch on the floors, chambers, and covers. Sterilize all frames.

Keep your bee suit reasonably clean. Your gloves should be washed and replaced regularly. If handling hives in another beekeeper's apiary, wear clean gloves. Do not reuse these gloves to inspect your own bees unless they have been thoroughly cleaned.

4. Regularly replace old brood frames every two to three years.

5. Do not feed bees honey from another source.

6. Inspect your bees regularly and handle with care to avoid trapping and squashing the bees.

7. In the case of foul brood, you are legally obligated to report it to your local bee inspector or the national advisory board.

8. Carry out a thorough disease inspection in spring and autumn and, if in doubt, always seek advice from an experienced beekeeper.

Time to harvest

There are a number of by-products to be gleaned from your hive if you want to maximize your return. The more specialty products from your hive, outlined in this chapter, include propolis, pollen, and royal jelly, but the methods of collection are a little more involved and require experience and technical know-how.

Honey

Lifting out that first frame of capped honey is a wonderful sight. It looks fantastic, it's ready to eat, and it's just like the honeycomb you buy! The dense weight of one frame filled with ripe honey has to be handled to be believed, weighing approximately 3–5 lbs. Your goal has been realized and the surplus honey is yours to harvest.

Achieving a good crop of honey is reliant on managing a number of factors, but the weather is the one thing we have no control over. No matter the variety of flowers available for the bees to forage, and the care taken with your hives, it is the unreliability of the climate that can govern so much. The wrong mix of weather, unfavorable temperatures and humidity at critical moments in the season can badly affect the nectar yield. If there is little nectar for the bees to extract, the honey flow will be feeble and your harvest may be smaller than anticipated. But if you can maintain a strong colony, are lucky with good spells of sunny weather coinciding with the flowering of local plants, providing the bees with a good honey flow, you will be on track towards reaping your well-deserved reward. You may also make a little money to cover your costs by selling your honey and wax and with some creative thought, a little profit, too.

The conversion of nectar to honey

There are many different types and qualities of honey and each will depend on the flowers from which the worker bees gathered the nectar. You will notice that the honey made at the beginning of the season, stored in your first super, will be a different color, probably clear to pale yellow, compared to the darker honey made later in the season. The taste and aroma will vary, too. These differences occur when the source of nectar available to the bees changes throughout the season.

Nectar is a sweet, watery solution collected from individual flowers—containing a variety of substances including sugars—that provides the bees with their major source of energy. It is only in the ripening of the honey during its time in the hive that it becomes thicker, transforming into a high-density food.

Sometimes the bees will gather honeydew, a sweet substance that forms on the leaves of trees and plants like dew drops. Honeydew is either exuded in globules from plants or secreted by insects like aphids. On a sultry summer's day, oak trees can be seen alive with bees gathering sweet, sticky honeydew.

If your hive is situated in a garden or close to other gardens, your bees will have a wider variety of local flora from which to source their nectar and pollen over a substantially longer period than if situated in the countryside or on farmland. If, however, the hive is positioned next to a major source of nectar, such as orchards and crops (particularly oilseed rape), the bees will collect nectar, at a considerably faster rate and produce large quantities of surplus honey in a much shorter amount of time.

Harvesting

TOP TIP
Harvesting at early dawn or dusk will help prevent any robbing of honey by other bees.

Cutting the comb and watching the ripe honey break through, oozing out into clear pools of glistening thick liquid, is a fantastic moment. That first taste of your own honey is a moment of sheer satisfaction.

You will know when that exciting moment has arrived and your honey is ready. The harvest is usually carried out at the end of the season, although it is possible to remove the honey when a super is full. Some postpone the harvest until late August to ensure the honey is fully ripened and of a good consistency.

Check that the honey you are removing is completely capped, because the bees will only cap the honey once it has reached the correct density. In this state the honey will survive for a long time without fermenting (see Inside the hive, p.49). If you are impatient and take unripened honey before it has been capped, it will be very thin and watery, prone to fermentation, and therefore no good for storing. In short, you will end up with an unnecessary and disappointing result.

If you only have a hive or two to harvest, begin the process by placing a clearance board with a porter bee escape in position under the supers you wish to harvest from (see Basic tools and equipment, p.22). Ensure your supers are well-fitted, and not slewed out of position, as any gap will be found and your honey robbed by other bees and wasps.

The porter bee escape will do its job of acting as a one-way door, allowing the bees down into the hive but blocking their entry back up again into the super. Left with this in place for at least twenty-four hours, the supers should clear, enabling you to take the honey with very few bees to contend with.

On a small scale, it is possible to harvest your honey without the use of a clearance board and porter bee escape. A frame of honey can successfully be cleared by gentle strokes with a soft brush to tease the bees away.

When you have taken the honey safely away from the hives and are inside with doors and windows shut to keep out any interested bees, you can start the extraction or comb cutting.

Methods of extraction

Squeezing the honey out of the comb is the most primitive way to extract the liquid from the wax, but this crude method is wasteful and therefore undesirable when there are other methods available.

Catalogs advertising beekeeping equipment have reams of pages with elaborate pieces of gadgetry for capping, extracting, and storing honey—so much so that the beginner can be edged into a slight panic.

If you are starting out in your beekeeping career on a budget, do not be duped into thinking you will need all these extraneous pieces of equipment. Simply cutting your comb into blocks, for example, is an easy, inexpensive, and attractive way to present your honey for use at home, and is always a popular honey product at markets and festivals.

Cut comb

Varying sizes of comb-cutters with matching comb containers are available. If you intend to sell your honey, it is best to use these tailor-made cutters because they will yield the correct size and weight. If you are cutting the comb for your own consumption, simply divide it into sections with a knife, slip each block of comb carefully into a tub, and clip on the lid. It couldn't be easier!

Some beekeepers will tell you to place cut comb in the freezer overnight to kill off any wax moth larvae that have embedded themselves in it. Any rogue element in the comb is usually fairly obvious though and so easily avoided. In any case, it really should not be a problem in a well-maintained hive.

Using a honey extractor

The honey extractor is, of course, a great piece of equipment if you want to produce beautifully clear, liquid honey. There are two types of extractor in which the combs can be positioned for extraction tangentially or radially, and in different sizes, from hand-operated tabletop extractors to larger, heavy-duty power-driven ones. Try and borrow your local association's extractor or that of a fellow beekeeper, which will give you some experience of using one before you decide which type to purchase for yourself.

The kitchen is probably the place where most of us have to carry out the job of honey extraction—unless you are lucky enough to have another suitable room with running water and wipeable floors and surfaces.

Before placing the frames of honey in the extractor, the cappings have to be removed to release the honey. This is done with a special uncapping fork or an uncapping knife, warmed to cut through the wax. You will need a deep tray (specialty uncapping trays are available) or large bucket, with a strainer ready to catch the cappings and dribbles of honey that will fall as you release the honey. The warmer the combs and the environment you are working in the more relaxed the honey will be, making it easier for it to slip away and run from the wax. Cut away the cappings by slicing as close to the surface as possible, and angle the frame so that the cappings will drop away from the honey into the strainer. If the strainer becomes congested with cappings give it a gentle stir to release the honey, allowing it to drip through. Those with a liking for mead can use the the honey that is left on the cappings to make the tasty liquor (see Recipes and ideas, p. 96). Otherwise, the dried cappings can be added to your wax harvest.

Place each prepared frame into the extractor until full and then begin to rotate. The centrifugal force will pull the honey away from the wax. Go slowly to start with to avoid the comb from breaking under the strain of the momentum and gradually increase the speed of rotation.

Keep an eye on the level of honey rising in the extractor. When you are ready to release the honey, place a strainer with not too fine a mesh (this only needs to catch the extra large pieces of stray wax) under the valve of the extractor. The honey needs to drip quickly through this into your strainer tank. Mini strainer tanks can be purchased and these work well for those managing up to three hives. The fine perforations in the mesh of these strainers catch the remaining particles, wax, and speckles of propolis and pollen. Of

NOTE
A stainless steel or food grade polythene extractor must be used if you intend on selling your honey.

course, you can construct your own device to do this job.

NOTE
Many prefer not to have their honey overstrained as the nutritional value of the unfiltered particles in the honey is regarded as extremely nourishing (see pp.80–84).

When you are satisfied with the quality of your filtered honey it is best to let it stand for at least twenty-four hours before bottling. This will allow any bubbles and residue to rise to the top of the tank.

The best way to deal with the wet frames you are left with after the honey has been extracted is to replace them in the hive for the bees to deal with. They will be more than happy to clean up the sticky mess and you will be left, once again, with drawn comb ready for the supers, giving the bees a head start.

Bottling the honey

Depending on the amount of honey you have extracted you may wish to start bottling it as soon as possible, or you may prefer to keep it in a tank for bottling as and when required.

You can choose from a variety of specially-made containers in which to store your honey, and these are readily available from suppliers of beekeeping equipment. They range from inexpensive plastic pots with snap-on lids to novelty pots to traditional screw-top glass jars.

For your own purpose you can use any jars you have managed to collect, but if you wish to sell your honey, the containers must be of a legal size and standard (see Selling your hive products, p.108). It is advisable to refer to the latest regulations set out by the USDA and the FDA that will explain all the regulations you need to meet to sell "at the gate." If you belong to your local association find out if they are buying the jars wholesale as this will obviously keep down the cost per unit for you.

Note: Be aware that because of honey's hygroscopic nature (meaning it will absorb water from the atmosphere), it must be kept in airtight containers to avoid fermentation.

Overheating, which can occur during the use of certain methods to warm the honey in the extraction process, can also affect the honey by caramelizing its flavor. The same can happen to honey over a longer period of time if the conditions it is kept in are too warm. It is worth taking the time to think about where you will store your honey to avoid spoiling which may affect its color and flavor.

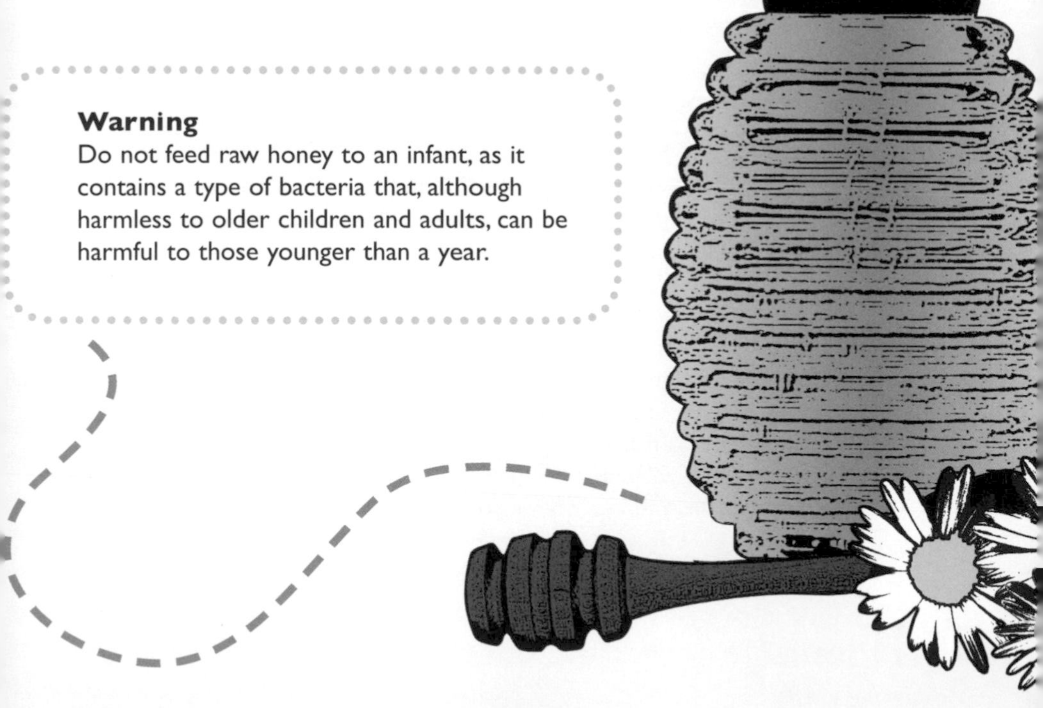

Warning

Do not feed raw honey to an infant, as it contains a type of bacteria that, although harmless to older children and adults, can be harmful to those younger than a year.

Beeswax

Wax is such a valuable part of the harvest, and bees were originally kept for their wax as much as for their honey. Before the Middle Ages, tallow (animal fat) was relied upon as the principal ingredient in candle making but it burns with a lot of smoke and has an unpleasant odor. The introduction of beeswax candles, which burn cleanly with a bright light and a sweet aroma, was a huge improvement and highly prized.

Before alternative forms of wax and lighting were available, beekeeping and monasteries were always closely linked due to the monks' need for wax and, of course, for the nourishing honey. A fine example of a monastery and its apiary still thriving today is Buckfast Abbey, thanks to the dedication of internationally renowned beekeeper, the late Brother Adam.

Beeswax is still an important commodity today and is widely used in furniture polish, cosmetics, pharmaceutical products, leather conditioner, the coating of sweets, and all sorts of crafts. A large amount is also reabsorbed by the beekeeping industry in the production of beeswax foundation for hives.

Laying the foundations

Bees could manage perfectly well without the foundation placed in the hive for them but it encourages them to start building up the comb—and just where you want it, too.

For a bee to produce wax it first needs to feed on lots of honey or syrup from a feeder; a significant amount of this is required (figures vary from 5.5–8.5 lbs. to make a single pound of wax). When the bee is well nourished, it will cluster together with other bees, which raises the temperature, and in a day or so it will start to secrete scales from glands

DID YOU KNOW
that it takes over half a million platelets to produce 1 lb. of beeswax.

on its abdomen. The bee will then chew on the platelet adding saliva and other secretions until it becomes malleable. It is then ready for the worker bee to position the shaped platelet, building up, piece-by-piece, the most amazing storage system.

Newly drawn comb is a truly wondrous sight, with each hexagon in delicate, fresh, pale wax, absolutely regular in build and scale. The hexagon is the most effective and economical of shapes to build and, because of its geometric and structural efficiency, it provides the bees with a durable storage system that offers maximum space for their honey and that of the brood.

Extracting the wax

Collecting beeswax is sometimes seen as a by-product of beekeeping, but it is equally desirable and as worthy of harvesting as the honey. If you can recover clean, reusable wax you have a product ready to be converted into polish and candles, or swapped for cash or new foundation.

Beeswax can be collected throughout the season by clearing away the unwanted brace comb, saving the cappings and surplus frames of comb after the honey extraction, and then melting the wax away from the solid honey.

There are various methods by which you can extract the wax. Some work better depending on the quantity you are working with and others produce a better quality of wax. Decide which one will best suit your needs.

Solar extractor

A solar extractor is an incredibly economic way of reclaiming your wax—all you need is the sun! If you are inclined, there are plenty of plans available showing you how to build your own. A solar extractor is basically a box with a double-glazed top. Inside is a titled tray for taking the comb and a container for receiving the molten wax, which then needs to be poured into molds and cooled for storage.

Relatively clean wax can be obtained with a solar extractor and strainer, but it may require further filtering. To do this, gently melt the wax in a bowl placed in a saucepan of warm water. Your cakes of wax may need rinsing before you do this.

TOP TIP

It is advisable to use rainwater on wax as your tap water may have lime in it that will react with the wax, causing it to become brittle and spoil the finish.

Cloth bag

Another inexpensive method is to place all your pieces of beeswax into a cloth bag (muslin or cheesecloth work well) that will allow the wax to melt through the bag while holding back the contaminents. Place a weight, such as a stone, in the bag with the wax bits and gently plunge it into a stainless steel pan of hot water. The water should be no hotter than 149°F otherwise you risk your wax losing its light honey color for a darker shade. It is helpful to use a jam thermometer or other temperature gauge when doing this. Let the bag sit in the water until the wax has had time to melt and leach through the bag, and rise to the surface. Remove from the heat source and leave to cool. When cold, the wax will solidify, enabling you to lift it from the water. The wax may still not be pure enough for candle making or molding and you may wish to put it through an additional filtration. The water bath is one method, described here.

Warning
Wax is very flammable, so never melt it over a direct heat source. Always heat it in a bowl over a saucepan of water so that the water, not the wax, receives the direct heat.

Double saucepan
Using the double saucepan system is another method suitable for filtering small amounts of wax. The wax is placed in a steel saucepan or heatproof bowl which is then placed over another saucepan of warm water. Add the pieces of wax to the warmed bowl and allow to melt. Strain the molten wax to clear it of impurities and let it filter through into a heat resistant container. A very fine wipe of washing up liquid will help to release your cake of wax when it has cooled.

Steam wax extractor
A steam wax extractor is a more expensive option but it will allow you to produce a cleaner wax in one go, probably only needing additional filtration if your wax is for show purposes.

Storing your wax

Wax can be stored in cakes or blocks of any size but if you allow your freshly filtered wax to collect in a bowl of cold water it will instantly solidify, leaving you with small pieces. Retrieve these from the water, place on a cloth, and allow to dry. These smaller, more delicate pieces are easy to store and quicker to melt when required.

Stored in cool, dry conditions, beeswax has an almost indefinite shelf life. Due to certain components coming to the surface, it may eventually get a pale powdery coating. This bloom is normal, though storage conditions can encourage it to develop, and has no damaging effect. It can be removed by warming the beeswax or by wiping it gently with a soft cloth.

Bee pollen

Pollen is a very fine powder collected by the bees from trees and plants to provide themselves with nourishment. It is one of the richest foods in nature, containing a wide variety of proteins, minerals, and fats, together with almost every vitamin.

The benefits of eating bee pollen are huge and many claim there is no greater health product available. It is not surprising, therefore, that it is regarded by many as a powerful "super food."

Pollen dust is made up of minute grains, or cells, that collect in a powdery form on the legs and body of the bee as it works its way in and out of flower heads. While performing this task, the bee also acts as a pollinator—essential for most plants as they cannot self-pollinate. The act of pollination plays a crucial part in a plant's reproduction cycle. After visiting several flowers, the bee will comb the pollen down her body with the brushes on her legs, moisten it with a drop of honey, and pummel it into pellets that are placed in the pollen baskets for the flight home. When the pollen baskets are laden they are clearly visible to the naked eye sitting on the bee's legs.

The pollen is taken back to the hive and stored in an empty or partly-filled cell in the comb close to the brood. It is mainly used by the nurse bees to feed the larvae. It is also used in the production of royal jelly, the staple food for the queen bee. Cells that have been packed with pollen are easy to differentiate from those holding brood or honey, showing up as a pallet of varying shades from gray to yellow to orange, blue, and scarlet. The color variance is because the honeybee will collect only one type of pollen on a single flight—if it has visited an apple orchard the pollen will be a chalky yellow, compared to the pollen collected from Horse Chestnut trees, which will be a deep red.

Pollen extraction

Not many beekeepers collect pollen as part of their harvest but some may gather enough to make pollen patties with which to feed the bees when there is a scarcity of pollen, sometimes in early spring. Pollen traps can be purchased and these are fitted at the front of the hive to catch some of the pollen as it brushes off the legs of the bees. The pollen caught at the hive's entrance is different to that already stored in the cells, which is known as bee bread. The bee bread will have had its composition slightly changed by the bees through the addition of a little honey, propolis, and other secretions. In this state, it is scientifically proven to have even greater health-giving properties than the dry pollen collected by a trap. Note that any collected pollen needs to be stored in the fridge or freezer.

You may find that after extracting your honey you are left with some pollen still in the comb. This can be saved by scraping out the pollen and the cell it is held in. If you can avoid cutting a hole right through the wax your comb will be ready for the bees to use again next season. Again, freeze the pollen for later use.

If you are interested in harvesting your pollen for commercial reasons, there are a growing number of traders and suppliers who can be contacted for their terms and conditions.

Propolis

Propolis is a resinous substance, variable in color (but commonly reddish brown), and very viscous. Collected from trees and plants, particularly from leaf buds, flowers, and the tree bark, the bees use it to seal unwante gaps in the hive. For most beekeepers, propolis, or bee glue, is a sticky nuisance that has to be contended with. When inspecting your bees, the various sections of the hive will need to be encouraged apart with your hive tool as they will have become glued together with propolis.

It is believed that this sealing of components makes the hive more defensible, stable, and waterproof, and prevents disease and parasites from gaining entry. It is an interesting fact that if a larger intruder, such as a mouse, manages to squeeze its way into the hive and subsequently dies, it will, if the body is too large to be removed, be mummified in a coating of propolis, preventing its decomposition from contaminating the colony.

The bees collect the propolis by chewing resin, a sticky substance found on trees and plants, until it is malleable enough to be placed in its pollen baskets for the flight back to the hive. Here, it will hand over its supply to other worke bees to use as required. Propolis is not only used as a sealant, but also as an antiseptic varnish for the inside of the hive to protect the colony from infection—in the same wa that the resin exuded by injured trees helps to seal wounds, defending the tree against attack from insects, bacteria, and fungal spores.

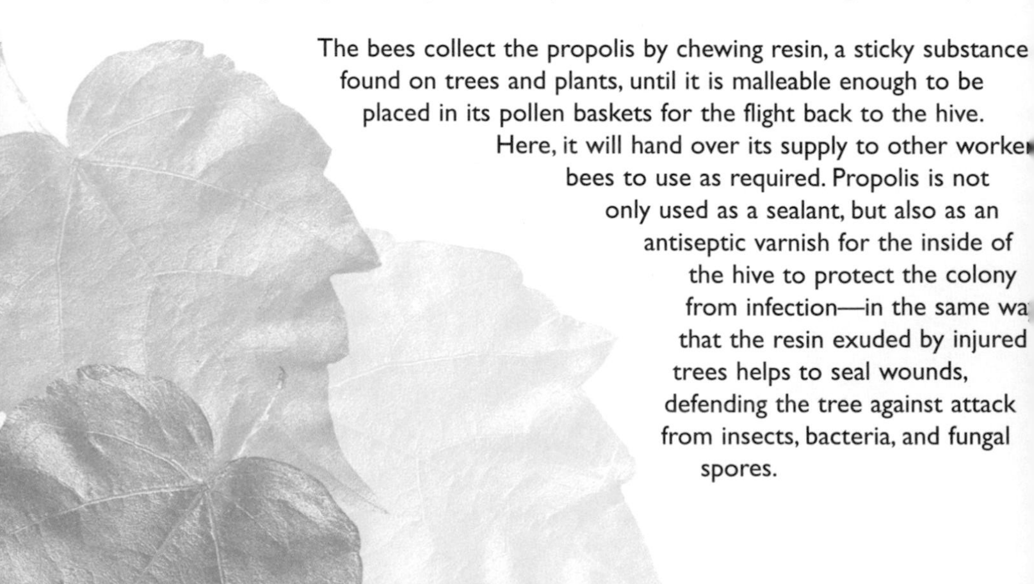

Health benefits
Due to its powerful combination of properties, acting as an antibiotic, antiviral, anti-inflammatory, and antifungal substance, propolis is now acclaimed for its beneficial, health-giving attributes.

Historically, the use of propolis is mentioned by the Ancient Greeks, Egyptians, and Romans as a remedy for a number of ailments including infections, swellings, and the healing of sores and skin diseases. It is also believed to have been used for mummification.

In more recent times, before the introduction of antibiotics, propolis was mixed with petroleum jelly and applied to dressings to successfully heal and disinfect wounds. Many of these natural methods were disregarded and forgotten with the launch of modern drugs. In recent years, there has been renewed interest in the use of propolis as a natural remedy, and it is also the subject of research in the conventional medical world.

Collecting your own propolis

Collecting your own propolis can be done with the help of a propolis screen placed under the crown board of your hive. When the mesh is covered in propolis, remove it from the hive and freeze it until the propolis becomes brittle. To release the particles of propolis, scrunch the mesh together and they will drop off with ease.

Royal jelly

Royal jelly is a truly special substance—a thick, creamy-white liquid, rich in vitamins, proteins, fats, sugars, and other components. It is a well-known fact that whether an egg develops into a worker bee or a queen bee is determined by the quantity of royal jelly the larva is fed—a potential queen bee is fed substantial amounts. The royal jelly is produced by the young nurse bees from regurgitated nectar that has been mixed with secretions from their hypopharyngeal and mandibular glands.

It is the extraordinary effect this high-energy food has on the developing queen bee, that it has led to it becoming so highly prized. It is believed by many to have outstandingly beneficial health properties, associated with increased energy, immune system protection, and general longevity.

Royal jelly preparations are readily available from health food stores and it is also commonly seen promoted for its goodness on the labels of soaps, cosmetics, and health supplements.

Royal jelly extraction

To produce royal jelly for the market, a lot of queen bees need to be reared, so those who farm bees for their royal jelly manipulate their colonies to produce lots of queen bees. The more developing queens there are the more royal jelly the worker bees need to produce to fill the queen cells. At the point of harvest, the beekeeper removes the larva from the pool of royal jelly it is lying in and then extracts the liquid for immediate freezing.

Selling honey products

By referring to the latest information set out by the regulatory bodies in your area that deal with food or trading standards, you can find out about the regulations you will need to meet in order to sell your honey products.

Recipes and ideas

Honey, wax, pollen, propolis, and royal jelly are found in a wide variety of products. In cookery, it has been used for thousands of years as a sweetener, in salad dressings, in sauces and marinades for meats, and in biscuits, breads, and desserts. Here are a few fun recipe ideas to try with your newly acquired honey and wax.

Cooking with honey

Substituting honey for sugar gives a longer shelf life to cakes, scones, and bread. Honey retains moisture more effectively than sugar, so foods baked with honey remain moist for longer than those baked with sugar. Using honey in baking where fruit is also used will enhance its flavor.

Honey, Banana, and Apricot Loaf

- ½ cup soft apricots, cut to 6 pieces
- 5 Tbsps runny honey
- 2 cups plain flour
- 2 tsps baking powder
- ¾ cup light Muscovado sugar
- 6 tsps butter, softened
- 3–4 bananas, about 1½ cups
- 2 large eggs

Grease and line the base of a 2 lb. non-stick cake pan. Preheat oven to 350°F.

Place the apricots and honey in a small saucepan and bring to a boil. Simmer for 2–3 minutes, stirring frequently. Allow to cool for 10 minutes.

Place the other ingredients in a food processor and blend until smooth. Remove the blade and stir in the honey and apricots.

Spoon the mixture into the cake pan and bake for one hour, covering the top of the loaf with a piece of foil for the last 20 minutes of cooking time. The cake is ready when a skewer inserted into the center comes out clean. Cool for 15 minutes in the tin then turn out onto a wire rack. Serve in thick slices, warm or cold.

Honey Scones

Preheat oven to 375°F.

Mix together the flour, baking powder, and salt. Rub the fat in well and then add the honey and fruit. Mix with a little milk until you have a light dough.

Roll out dough to a depth of about a half inch and cut into rounds. Place in the oven and bake for 12–15 minutes.

Honey and Sesame Squares

A healthy and satisfying snack at any time of the day. The combination of sesame seeds and honey is particularly tasty.

Preheat oven to 350°F.

Combine all the ingredients and press the mixture into an 8-inch shallow square tin.

Bake for 30–35 minutes until golden brown. Allow to cool in the tin, then cut into squares.

Honey Buns

Delicious at tea time!

Grease bun tin or use paper cake cases placed on a baking sheet. Preheat oven to 375°F.

Sift the flour and salt into a mixing bowl. Rub in the butter or margarine until the texture resembles fine bread crumbs. Mix in the sugar.

- 1 cup self-rising flour
- 6 Tbsps butter or margarine
- 2 Tbsps caster sugar
- 1 egg
- 2 Tbsps liquid honey
- 2 Tbsps milk
- pinch of salt

Whisk the egg in a separate bowl and add the honey and milk. Mix until well blended, then combine with the dry ingredients and beat until smooth.

Divide the mixture evenly in the bun tin or paper cases and bake for 15–20 minutes until it rises. Place on a wire rack to cool.

Makes 18–20 buns.

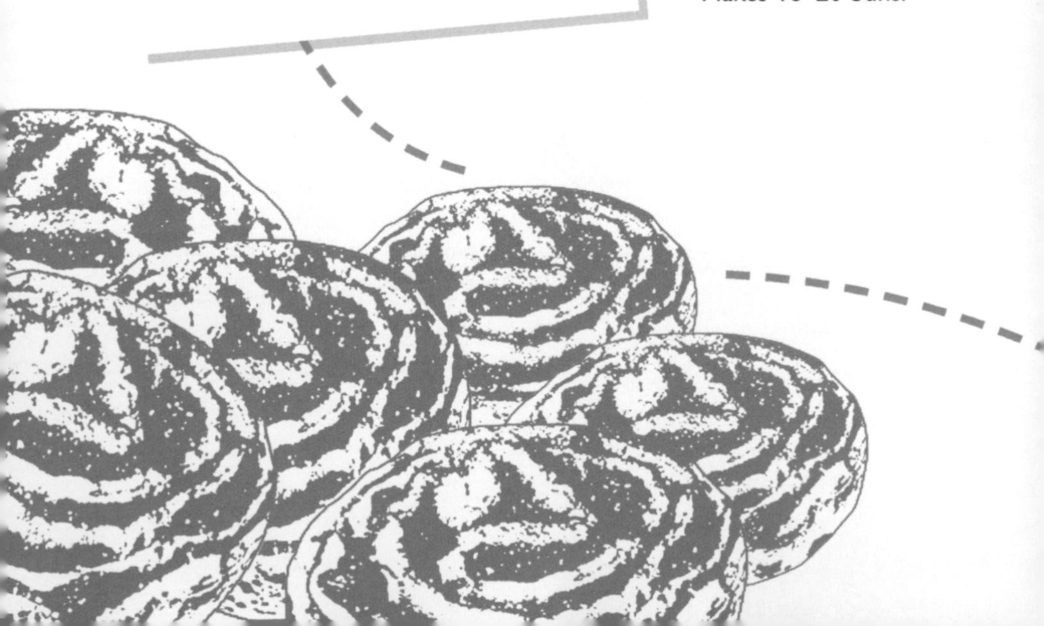

French Bread and Honey Pudding

A great way to use up a stale baguette. It makes a wonderful base for this new take on a traditional bread and butter pudding.

- 2½ cups 2% milk
- 1 vanilla pod, split
- 4 eggs
- 2 Tbsps caster sugar
- 1 Tbsp and 1 tsp unsalted butter, softened
- 16 slices French bread (baguette)
- 2 Tbsps and 2 tsps Set honey
- ⅓ cup ready-to-eat dried apricots, chopped
- ⅓ cup and 1 Tbsp raisins
- ⅓ cup and 2 Tbsps pecan nuts, roughly chopped
- 1 tsp grated nutmeg

Preheat the oven to 350°F.

Place the milk and vanilla pod in a saucepan and bring to a boil. Leave to stand for 10 minutes, then remove the pod.

Beat the eggs and sugar together, add the milk, and whisk together to make a custard. Thinly butter the bread and spread with honey.

Combine the apricots, raisins, and pecan nuts and sprinkle half over the base of a shallow 2-pint dish. Arrange the bread slices in the dish so they overlap slightly. Sprinkle over the remaining fruit and nut mix and cover with the custard. Dust with nutmeg and leave to stand for 15 minutes.

Place the shallow dish in a roasting tin, adding sufficient boiling water to reach halfway up the sides of the dish. Bake for 40–45 minutes, until set and golden.

Marlenka, or Mednovik (Honey Cake)

Mednovik is a delicious torte made of layers of honey cake with a creamy filling. Popular in the Czech Republic, it is beautifully light and moist to eat.

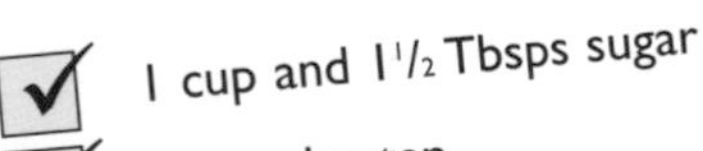

- 1 cup and 1½ Tbsps sugar
- 2 eggs, beaten
- 4 Tbsps butter
- 2 Tbsps honey
- 2 Tbsps baking powder
- 3 cups all-purpose flour
- Cream Filling (recipe on next page)

Preheat the oven to 375°F.

Lightly flour a baking sheet. Prepare five sheets of wax paper cut into 8-inch circles. Grease the paper lightly.

In a small bowl, combine the sugar and eggs, then set aside. Melt the butter in a large saucepan over low heat. Add the honey, the egg/sugar mixture, and the baking powder, and stir constantly until blended and foamy. Remove from the heat and stir in enough flour so that the dough is not sticky.

Separate the dough into five equal portions and place each one onto the wax paper circles; cover each portion with plastic wrap to keep warm. Using a floured rolling pin, roll one section into a round one-fourth inches in depth and place on the baking sheet (remove wax paper at this point). Bake for 3–5 minutes until barely golden but not brown. Remove from the oven and transfer rounds to a wire rack to cool. Repeat with the remaining four sections of the dough, reflouring the baking sheet if necessary.

Prepare the cream filling (see below). On a large serving dish, alternate five layers of cake with the cream filling, spreading the cream liberally between each cake layer. Crumble the final cake layer into small pieces and sprinkle over the last cream layer. Let the cake sit for a day before serving.

Cream Filling

1 can (14 oz.) of sweetened condensed milk

3 eggs, beaten

2 Tbsps honey

4 Tbsps butter

Cream Filling
Place a large saucepan over medium heat. Combine sweetened condensed milk, eggs, honey, and butter, stirring constantly. Gradually bring to a boil and maintain heat until the mixture thickens. Remove from the heat and allow to cool.

Roasted Honey and Spice Nuts

A great way to turn a plain bag of mixed nuts into a special treat.

-  2 Tbsps butter
-  2 Tbsps runny honey
-  1-2 tsps mild chilli powder, (according to taste)
-  ½ tsp salt depending on preference
-  freshly ground black pepper
-  1½ to 2½ cups mixed unsalted nuts (choose from almonds, cashews, pecans, hazelnuts, walnuts, or peanuts)

Preheat the oven to 300°F.

Place the butter in a saucepan and heat gently, until melted. Stir in the honey and season with chilli powder, salt, and freshly ground black pepper.

Add the nuts and stir until thoroughly coated with the honey mixture. Transfer to a shallow non-stick baking tray and roast for 20–25 minutes, or until glazed, stirring two to three times.

Allow the nuts to cool. They can be eaten warm or cold and will keep in an airtight container for up to a week.

Honey Mustard

- 1 1/2 cups white mustard seeds
- 1 1/4 cups white wine vinegar

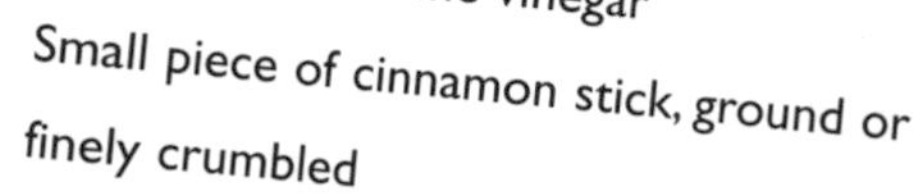

- Small piece of cinnamon stick, ground or finely crumbled

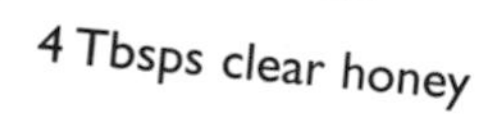

- 4 Tbsps clear honey

Place the mustard seeds, vinegar, and cinnamon in a bowl, cover, and leave to soak for 24 hours.

Grind the soaked mustard seed mixture together with the honey until it becomes a stiff paste (a mortar and pestle is ideal for this, otherwise use a food processor), adding a little more vinegar if required.

Decant the mixture into jars. Cover and seal. The mustard will keep in the fridge for up to a month.

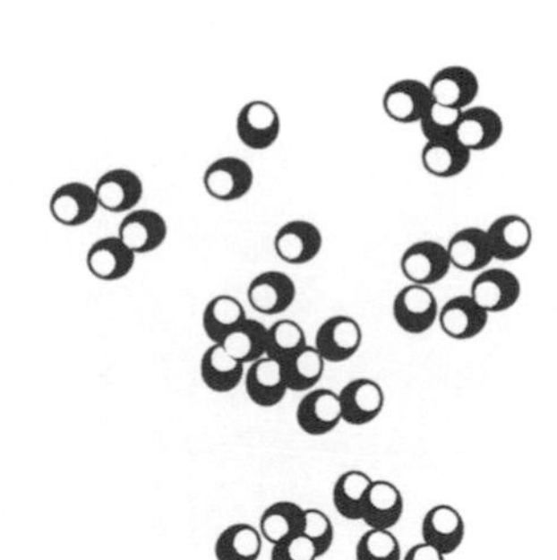

Honey and Mustard Marinade

Honey works well in marinades for meats such as duck, pork chops, or chicken and is easy and quick to make. To achieve an evenly cooked finish, wipe off the marinade and pour it onto the food half-way through cooking to produce an appetizing sheen.

Mix 2 Tbsps of grain or French mustard with 2 Tbsps of clear or runny honey and the juice and zest of one lemon.

Mead

Mead, made from fermented honey water, is one of the most ancient of alcoholic drinks. A mix of honey, water, and yeast known as "must" is the base ingredient. Once the must is bottled and has fermented for a year, it becomes mead.

- 1 gallon water
- 2 cups and 2 Tbsps honey
- 2 cups sugar
- 2 egg whites
- Peel and juice of 1 lemon
- 4 3/4 tsps of Champagne yeast (type of yeast that complements the flavor of mead)

When made well, mead can be a superior and fortifying drink. There are many recipes for it and, depending on the type of honey used, the flavor can vary enormously. In addition, experimenting with fruit, herbs, and spices, and blending meads, can make for some interesting tastes to savor. By adjusting the amount of honey or the type of yeast used, you can make the mead sweet, semi-sweet, or dry.

It was customary after harvesting the honey for nothing to be wasted; the dregs left in the cappings that can't be separated from the wax, and are not usuable for anything else, serve as the basis of mead.

Honeymooners' delight

The term "honeymoon" is alleged to have come from the ancient Norse custom of newly weds drinking mead for a whole month (or "moon") after their wedding to enhance their fertility.

There are a huge number of mead recipes to choose from, but here is one to experiment with—a simple, tried-and-tested recipe from a veteran beekeeper—and it's easier to brew than beer.

Blend the water, honey, and sugar together. Beat the whites of the two eggs to a froth and add to the mixture (this helps to give the mead a clear look and does not affect the flavor).

Boil the liquid for as long as any scum rises. When lukewarm, add the juice and peel of the lemon together with the yeast and let it ferment.

Allow the mixture to sit and "work" before bottling.

Medicinal

The method by which a bee produces honey, ingesting and regurgitating, means that by the time we get to eat it, it has already been predigested. This allows our bodies to process honey quickly, making it an easily digestible food—and excellent for anyone with a weak digestive system. Eating honey is also a speedy way to get an energy boost.

Honey and hot water

A cup of hot water with a generous spoonful of honey stirred in makes a delicious drink on its own.

Honey Hot Water and Lemon

Honey stirred in to a mug of hot water with freshly squeezed lemon juice added to taste is a popular remedy for a sore throat and chesty coughs and helps with insomnia, too. It is also the perfect drink to start the day, aiding digestion and cleansing the bladder and kidneys.

Heather Honey Hot Toddy

Heather honey mixed with a little whiskey and topped up with boiling water relaxes the symptoms of the common cold.

Honey Flu Remedy

A remedy to help alleviate bronchitis and flu. Slice a 6-inch ginger root and place it in a non-aluminum pot with about three cups of fresh water. Cover the pot tightly and bring to a simmer for about 20 minutes. Remove from the heat and add the juice of half a lemon, a pinch of cayenne pepper, and honey to taste.

Honey Poultice

Applying honey to minor cuts and abrasions will help to draw excess fluid from the injured area and reduce swelling. The hygroscopic and antisceptic qualities of honey help to prevent infection. Dab the honey onto the wound and cover with a clean dressing.

Honey tea for longevity

To stave off the ravages of time and keep the skin soft (and perhaps help to increase one's lifespan!), make a tea with four spoonfuls of honey, one spoonful of cinnamon powder, and three cups of water. Drink one cup 3–4 times daily.

Warning
The above remedies are not meant to be used in place of orthodox treatment. Always consult a qualified doctor before embarking on any treatment.

Eating Local Honey

By consuming honey that has been produced by an apiary in your area, you will be ingesting local pollens which is believed to, over time, help to build resistance against allergies and hayfever. The honey should be raw, not heated or processed.

Warning
Never feed raw honey to an infant, as it contains a type of bacteria harmful to those younger than a year.

Body care products

Many creams and lotions for the face, hands, and body contain honey, which is well known for its nourishing, astringent, and antisceptic qualities.

Hand Softener

Honey, egg yolks, and sweet almond oil mixed together will soften and smooth the skin.

Soap

Honey can be added to any basic, cold-process soap recipe. It adds a delicious and restful aroma to the soap, with a clean, fresh appeal.

There is a great variety of soap molds on the market to choose from. Molds depicting bees, skeps, hives, and comb are available through suppliers of beekeeping equipment.

Hand cleanser

A scoop of honey rubbed into the hands is a great cleanser and will get rid of fairly stubborn marks without leaving the skin dry and rough.

Honey and lavender milk bath

Grind the lavender flowers into a powder. Place the lavender powder, milk, and honey into a bowl and whisk together. When thoroughly blended, pour into a clean, lidded jar and keep refrigerated. Use within a week.

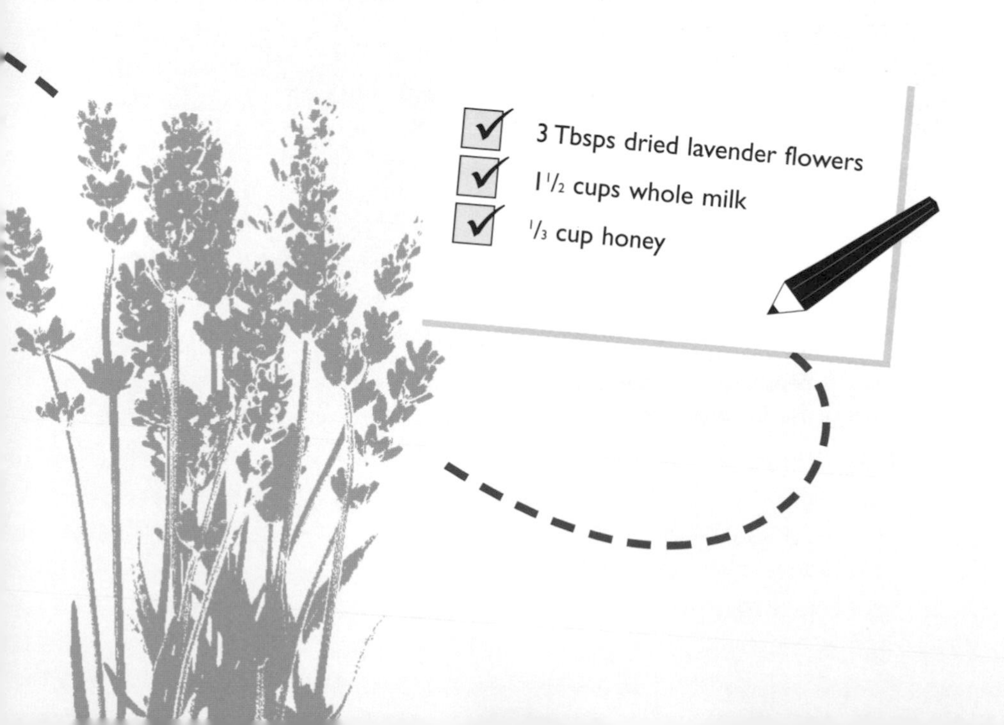

Beeswax products

Making your own beeswax foundation

Beeswax foundation is readily available cut to size, wired, and embossed and the majority of beekeepers will use this. However, you can make your own with a wooden sheet wax tray, some silcone spray, foundation matrix, and a ready supply of liquid wax.

The method for this is basically pouring enough liquid wax into the dampened wooden tray (spray with the silicone if required to prevent sticking). Roll it around to cover the surface of the tray, letting any surplus run out of the small gap in the tray wall. Using a blunt knife or spatula, ease up the wax at one corner and gently peel away the sheet of wax. Place it between the plastic foundation matrix and gently apply a roller to make the honeycomb impression in the wax.

If you find yourself with more wax than you know what to do with, you can arrange a straight swap set up by one of the major suppliers of beekeeping equipment. You deliver your clean wax to the factory and in return they will provide you with a fair swap of beeswax foundation ready to use in your hive.

Wax Molds

If you aim to harvest and extract the wax from your hives, special molds are available for producing 5-lb. bricks of beeswax in preparation for selling, or perhaps making your own polish. Smaller ingots of wax weighing 1 oz. can also be created. Other molds depicting patterns of bees, honeycomb, and hives are available, ready to turn your wax into attractive and beautifully scented objects.

For fun you can try using your own small glass pots, cups, and dishes. The safest way to melt the wax is in a water bath, a dish over a pan of hot water. Break up or grate the wax into small pieces, gradually adding to your bowl as it melts.

The molds must be clean and a little olive oil rubbed inside them will ensure no sticking occurs. To ensure the gradual cooling of the wax, have the molds warmed ready for pouring in the liquid wax. The best way to achieve this is to place the molds in a pan of hot water. If the wax cools too quickly by pouring it into a cold mold, you risk ending up with cracks in your wax models.

Lift the bowl of liquid wax from the heat of the pan and wipe the underside dry. If drops of water drip into the molds they will pit the wax.

Pour the wax slowly into the molds, pricking any air bubbles that form. When the wax has cooled enough to form a crust add more warm water to the pan that the molds are sitting in so that the wax is covered. Wait for the water to cool off and, when the wax is cold, it will pop up and float out of the mold.

Glossy Beeswax Polish

Recipes for beeswax polish vary widely in their ingredients, but this simple method makes a polish that will give a beautiful sheen to your furniture. Carnauba is a vegetable wax which comes from the carnauba palm and is available from most DIY stores. Its properties will help to harden the wax and provide a glossy finish.

You will need:

Beeswax
Carnauba wax
Turpentine

Beeswax is highly flammable so take care and melt your wax in a bowl over a pan of hot water.

Gently melt three parts beeswax with one part carnauba wax. It is easier to melt your wax if it is grated into finer pieces. Remove from the heat and stir in three parts turpentine. Polish tins of varying sizes, together with personalized labels, are available from specialty stores.

Hand-rolled Beeswax Candles

Rolled foundation is probably the easiest of candles to make, unlike the molded or dipped candles that are more time-consuming and require a few pieces of equipment. The rolled beeswax candle, with its honeycomb imprint, is a popular choice with the public. They look good and give a lovely soft light as they burn, while filling the room with the gentle perfume of fresh beeswax and honey.

You make these candles with your own prepared wax or with beeswax sheets available via your regular mail-order supplier or local craft store.

To make a rolled candle you will need a sheet of beeswax and, for ease of modeling, it needs to be pliable. If your wax is too brittle to be rolled, try warming it or relaxing it by passing it over a pan of steaming water. It is also helpful to do your candle making in a warm room.

With the wax ready to roll, place a sheet of wax paper on the surface you will be rolling on. This will keep your wax free from pitting.

At one end of the wax, place your wick so that you have about an inch sticking out over the wax at either end. Fold enough of the wax over the wick to hold it in place.

With care, gently continue to roll the wax sheet with equal pressure along the line of the candle, keeping each end as neat as possible. When all the wax is rolled and the candle formed, press the edge of wax into the body. This should be easy to do if your wax is warm.

Choose which end you wish to present as the top of your candle and trim the wick to ¼ in.; any longer than this can make the candle smoky. Cut the wick flush at the base of the candle.

If you have rolled your candle with raw wick, (or even if you are using waxed wick), dip the wick into some melted wax to prime and give it a professional finishing touch. Simply take a little of the beeswax and melt it in a dish over a hot saucepan of water, then dip the wick of the candle and stand to dry.

Experiment with different sizes and colors of wax to make larger, longer, or shorter candles.

Selling your hive products

Having your own delicious honey to enjoy and share with family and friends is a great feeling—it's a wonderful, natural product and a delight to give and receive. If you are fortunate enough to end up with more honey than you need, then you could consider selling it to help fund your bee maintenance for the following year. Honey is a popular food and fresh, local honey, in the jar or on the comb, is very desirable. The consumer has always been aware of the health-giving properties of honey, today more so than ever, with the increasing scientific evidence of its healing properties. Pollen and propolis are also popular as they are also seen as health-boosters.

If you have harvested and filtered your beeswax and have surplus cakes or bricks, these will also sell to people wishing to make their own polishes. Better still, if you have been practicing the art of candle making, or molding your wax into fun objects, such as mini hives, these can also sell well.

Those with larger apiaries can be seen selling their goods at country fairs and agricultural shows and the more substantial producers of honey may have the option of marketing their products through co-operatives and packers.

If you find yourself in the position of being able to attend a farmers' market or local festival to present your products, it may help if you can offer your customers a few of the amazing facts about honey, as well as tips on how to use it other than just on bread—delicious though this undoubtedly is! Try offering an easy recipe or two to inspire. Many people who buy cut comb for the first time are puzzled as to how to eat it. Let them know that the wax is totally edible (eating it on hot toast is probably a favorite). For those wishing to minimize the intake of wax, another way is to take a piece of comb and give it a couple of squeezes on the plate with the knife; the honey will easily ooze away from the comb. Those in the know enjoy honeycomb for its unfiltered properties, benefiting from eating particles of pollen and propolis.

Another way some beekeepers manage to sell their honey is by finding a suitable local shop to present their jars of honey. If people read that the honey is from a local source it inspires the customer and is an added incentive to make a purchase.

Before you rush into selling your hive products there are a number of things to consider for successful marketing. Gaining a good reputation for high standards of hygiene, together with attractive and accurate presentation, will help you to cultivate good customer relationships.

The following information offers some practical guidance for packing and labeling your honey, helping you to understand and comply with the regulations relevant to the way you label and sell your product, with a few selling techniques, too.

Packaging your honey

As with all food products there are specific rules and regulations to follow when packing and selling your hive products. It is an offense to falsely describe or promote food, so a number of laws are practiced to protect the consumer against misleading information or pictures being used on food labels. Help and advice can be obtained by getting in touch with the trading or food standards agency in your area and of course, the beekeeper's association in your country.

Jars and pots

Use commercial honey jars, bottles, and purpose-made containers for cut comb. Not only will it present your product in a professional manner, but they will be of the required legal size and weight, enabling you to price each item accurately.

Labels

The information on your honey pot label needs to follow a few guidelines. You will have to describe the honey you are presenting and label it as being "honey," "honeycomb," "chunk honey" (which is honeycomb placed in clear runny honey), or "baker's honey," a description used when the honey is only suitable for cooking purposes. The word "honey" with an accurate description such as "pressed honey" or "heather honey," can be used, but note that you cannot use a specific flower or blossom name if your honey has not come wholly from that source. You can also use the word "honey" with a description referring to the area or location from where the nectar was collected. The description must be unambiguous, so as to avoid any potentially misleading information.

Batch numbering
It is important that your honey is identified with a batch or lot number so that its provenance is traceable. As you harvest your honey, a record will need to be kept as to which colony produced which batch.

Address of producer
Your name and address should appear on the label.

Best before date
A "best before" date must be given and a period of about two years is acceptable, although honey does keep for much longer.

Country of origin
This must be displayed separately from your address. In the United States, it can either say "Product of the US" or "Made in the USA."

Weight
A metric weight indication is required.

Label illustration
Be careful when enhancing your label with a picture. It must not show a specific flower—clover, for example—if the honey has not been predominantly collected from that flower.

Once you have worked out your preference for packaging your honey, do not let the product down by offering badly strained honey. This will be off-putting to the majority of consumers who are increasingly demanding and health conscious. Jars and pots must be well sealed, clean, and without a hint of stickiness.

Seasonal reference guide

For a lot of new beekeepers, spring will probably present itself as the time to take on the first hive. By attending one of the short courses that are run by local associations through the winter, you will probably find yourself, by spring, in the middle of an apiary gaining practical hands-on experience and guidance from experienced beekeepers.

Spring

The exact point at which spring arrives varies greatly depending on where you are located but in general, when the worst of the winter is over, the queen will start to lay again. With a new brood underway, the bees will need to call on their stores of honey and pollen. Unless exceptionally fine weather occurs they will not be leaving the hive to forage just yet, so it is advisable to check that they have plenty of stores.

Preliminary inspection

Even during the spring your bees may be in need of extra sustenance because stores are used more quickly as the hive gets back into action. Take time to observe the hive and its comings and goings. Are the bees already flying and returning to the hive laden with pollen? If they are, this is a good sign that the colony is intact and the bees are well nourished.

If you are unsure what is going on in the hive then take action, first by checking that the bees still have enough stores to start building up the colony. To avoid opening the hive, which is undesirable if the weather is cold or windy, you can check by "hefting" the hive to determine how light or heavy it feels and calculate whether it has become low in food stores. To do this with a degree of accuracy, you will have had to record the weight of the hive with its winter stores in late autumn. This can be done easily with a spring balance hooked under one side of the hive, lifting it to record its weight. Repeat on the opposite side and then add the two

weights together which will give you an idea of the total weight of the hive. Obviously, when you carry out the weighing in the spring and your readings are drastically reduced it is sensible to help the bees for a few more weeks by adding syrup or candy for them to feed on. You can make these for yourself, although the candy method requires a lot of boiling and stirring and ready-made bee candy can be bought from suppliers of beekeeping equipment. Baker's fondant is a good alternative and used by many. Bees also need pollen as a source of protein and there are recipes for making an artificial pollen if for some reason there is a shortage of natural pollen in the spring. Again, a lot will depend on where you are located as to whether this may become a concern.

Candy feed

2½ cups water

12 cups sugar

1 tsp cream of tartar if available

Boil, stirring continuously, until all the sugar is melted. Simmer for ten minutes and then allow to cool down to roughly 120°F. Stir the mixture to thicken it and decant into shallow containers.

Syrup feed

4 cups sugar

2½ cups water

Dissolve the sugar in the water at a ratio of 2:1. Use water from the tap, stirring occasionally until completely dissolved. A thinner syrup for emergency feeds in the spring is made with sugar and water at a ratio of 1:1.

Artificial pollen patty

- 3 parts soya flour
- 1 part dried yeast
- 1 part skimmed milk powder

Mix ingredients. Place the patty on top of the brood nest.

If you have collected pollen from a healthy hive in the previous season you can include this in your patty:

Pollen supplement

- 3 parts soyal flour
- 1 part dried yeast
- 2 parts pollen
- Honey for moistening

Bind ingredients together. Place the patty on top of the brood nest.

Water supply
Make sure there is water close to the hive as the bees will need it regularly in order to dilute and break down their stores. If you want to avoid them visiting your neighbor's pond, provide a special supply by placing a large dish or bowl filled with water nearby with a few protruding pebbles and mossy sticks in it for the bees to land on.

If your bees are able to break down their winter stores, and receive extra supplies of syrup or candy in the early spring, it will help them enormously, enabling them to quickly draw new comb and stimulate the queen into laying. Lots of new brood needs to be brought in the spring in order to establish a strong colony that is ready for action when the new season's honey flow commences.

Quick check inside the hive

If you are unsure about how well your bees have survived the winter, a very quick look inside can be done if a warm, breeze-free day presents itself. Before dismantling the hive, send a few gentle puffs from your smoker into the entrance. Remove the top sections including the queen excluder. Take out one of the outer frames and place a cloth over the brood box.

Gently ease the frames apart enough to view the brood without lifting from the box. As you approach the middle of the nest the state of the brood will become apparent. Check to see if there are eggs and sealed brood. You should also see stores of honey and pollen on the outer frames and around the brood.

If there is no sign of brood underway or if only drone brood (which has a much higher, protruding cap than the worker cells) is being produced, you will have to ascertain whether the queen is failing or if she has died during the winter. As you cannot begin manipulating your hive until the weather is much warmer, you have time to glean the opinions and advice of others; you could introduce larvae from another hive from which the bees have a chance of rearing a queen, or introduce a new queen.

Full spring inspection

Resist the urge to fully inspect your hive until the weather has properly warmed up, reaching at least 60°F. You will know when the time is right without even reaching for your thermometer. If you are comfortably warm in your shirt sleeves, it should be suitably warm enough for making your first inspection without fear of chilling the brood. Think through your inspection before you start so that you know what you are looking for and how you must react. Watching the bees and their progress in the hive is fascinating, but do try to resist opening up the hive more than you need to as each inspection does agitate and cause stress to the bees.

Signs to look out for

First, take a look around the vicinity of the hive and check for any dead or sluggish bees on the ground. Streaks of brown on the outside of the hive may indicate there is trouble with dysentery (see Hive maintenance, p.64). Clear away any debris around the hive as well as tall grasses that could interfere with the entrance.

When you open the hive, do it as gently as possible to avoid pulling out frames that have stuck together. Scrape off the unwanted brace comb and gently clean the queen excluder to prevent a build-up of wax and propolis Check it is in good condition—not stretched or buckled. Remove old or damaged frames without disturbing the pattern of the brood nest.

Clean and check the hive floor for signs of disease, particularly varroa (see Hive maintenance, p.63), taking all debris away with you.

If in doubt, ask for help

If you are unsure about spotting signs of disease or about the welfare of your bees in general, do seek advice from an experienced beekeeper who will be able to remove any anxiety.

This may be the first time you have checked the brood nest since last summer, but if you have already made a preliminary inspection (see p.112), now is the time to take action on your previous findings.

The important thing to look for is evidence of a laying queen. The queen will have been laying since early spring so you should see eggs, larvae, and sealed brood. The cappings should be flat, not sunken, and without holes. If all the sealed brood is domed and in a regular pattern then it is likely the queen is failing, laying unfertilized eggs which will produce drones. Another possibility to look out for in the brood pattern is irregular cells with sealed drone brood, which may be evidence that a worker bee (an unfertilized female) is laying. Carefully inspect the brood frames checking for the queen. If you cannot see her and you have another hive from which to take a frame of healthy, disease-free brood place it in the middle of the brood nest and see if the bees raise queen cells. The alternative is to acquire a new queen, which can be delivered to you by mail from a supplier of beekeeping equipment. You will have to decide what measures to take, but if in doubt, always seek advice from an experienced beekeeper.

As you become more experienced and acquire more than one hive you will have the option of uniting colonies in order to strengthen the stock.

Marking the queen

Taking time to find your queen and mark her thorax with a dab of color not only lets you spot her more easily (for the moments when you need to find out whether she is still in the hive), it can also indicate how old she is. An international code of colors is used to identify the year in which a queen was marked.

If the year ends in:
0 or 5 the color is blue
1 or 6 the color is white
2 or 7 the color is yellow
3 or 8 the color is red
4 or 9 the color is green

Special marking pens, or paints, can be purchased.

Spring is a good time to mark your queen. She will be easier to find as the colony will be slower and reduced in numbers. To hold the queen in position for this procedure you will need a press-in cage, a simple piece of equipment that traps the queen on the comb when you have found her. It is critical that you do not damage the queen when you do this, so make sure you have plenty of time to complete the task—you will need a steady hand!

Clipping the queen

There are experienced beekeepers who, as part of their swarm control measures, clip the wings of their queens. This renders the queen unable to fly and so lead a swarm. If she does manage to leave the hive she will fall to the ground, where the bees will quickly follow and cluster around her.

Adding supers

Mid-spring is the time to put your first super on as there can suddenly be a spell of warm, sunny weather, together with a profusion of flowers for the bees to start gathering from. Be on your guard, however, as your bees are still in danger of starvation if the weather is unfavorable. The bees must have enough stores to feed on throughout the year. Any shortage will hinder the growth of the colony and, as a beekeeper, you will want the strongest force possible to make the most of the honey flow at its peak.

Be prepared for the honey flow to increase by making up frames and fitting the foundation in advance (see Setting up your hive, p.37). If the honey flow is good, the bees will quickly need more space to draw new comb. If they have nowhere to go in order to expand their stores, they will become congested and frustrated in the hive which can lead to swarming. It is essential to keep an eye on the speed at which the frames are filling up so that you can judge when a new super needs to be added. A good tip for doing this is to place a sheet of newspaper between your first super and the new one you are placing on top. By separating the two, you are not only keeping the hive insulated, but also allowing the bees to decide when they want to start on the new frames. When the bees are ready to extend their honey stores they will easily chew through the newspaper.

Summer

By late spring the beekeeper's objective is to encourage the colony to increase in size, boosting the chances of a good honey surplus. To achieve this, swarms need to be avoided and so regular weekly hive inspections are essential.

Swarm control can be the bane of beekeeping, especially for those in their first year or two. The mere mention of swarming in beekeeping circles is enough to raise eyebrows—and a few temperatures, too! However, with a few preventative measures and a couple of seasons' experience, the problem will become less troublesome.

Swarm control

From mid-spring through to summer your hive will need to be inspected on a weekly basis, certainly with no longer than nine days between each inspection, if you want to be able to take action to prevent swarming (see Inspecting your hive, p.56).

Check that the brood area has space for the queen to lay eggs. Introducing new comb into the brood box will alleviate the chance of congestion that can produce unnecessary swarming.

Catching a swarm

If your hives are located in your garden and you happen to be at home, you may see the moment the bees swarm. Otherwise, you may be completely unaware that the event has taken place until the next time you check the hive (see Hive maintenance, p.57). If this is your first season, it is highly advisable to get the help of an experienced beekeeper to show you how to catch the swarm.

Be prepared by having ready:

- Full personal protection (your bee suit)
- A smoker
- Straw skep or strong cardboard box with no lid
- Hessian sacking or cloth to wrap the skep or box in
- A small piece of board that your skep or box can sit comfortably on
- A soft brush for gently persuading the bees into the skep or box
- An hour or two to spare!

You may also need a ladder and pruning shears to cut away branches and twigs for a clear pathway to the swarm. If you are collecting a swarm from a neighboring property, make sure that you ask for permission before cutting trees or hedges.

How to catch a swarm

1. Place your piece of board on the ground and cover with a cloth, spread out beneath the swarm.

2. Take some time to judge in which position you will feel most safe and comfortable. With one arm outstretched, carefully place the skep or box as close and as directly underneath the cluster of bees as possible.

3. With your other hand, hold onto the branch the bees are clustered on and give it a firm shake or a knock with a strong stick to dislodge the swarm. The majority of the clustered bees will drop into the box below. If the swarm is fairly large, be ready for the weight of it when it lands in the box as they can be quite heavy.

4. With most of the bees in the box, quickly turn it upside down onto the cloth (see step 1). Prop up one end with a stick or something similar that will allow the stragglers—those flying around and those still left on the

branch—an entrance through which to find their way into the box. If the queen was caught with the main cluster, the other bees will soon sense where she is and gravitate towards her, into the box.

5. Allow the bees to gather and settle in the box for a least an hour, possibly waiting until the end of the day before returning to be sure all the bees have regrouped. Pull the corners of the cloth together and wrap over the box. You can now transfer the swarm to its new home.

6. You may want to keep your swarm or you may be happy to pass it on to another local beekeeper to add to their apiary. If you intend to keep it, have a hive ready and prepared with a few frames of new foundation and, if possible, a frame of old brood comb to get the bees interested.

The bee walk
For a fascinating sight, allow your newly collected swarm of bees to walk into the hive. Place a plank of wood angled from the ground up to the hive entrance with an old sheet draped over it. Position the box containing the bees at the base of the incline and eventually the bees will start to walk, in a line, up the slope and into the hive.

After transferring them to the hive, leave the bees alone for a couple of days and then introduce a feed of syrup that will encourage them to quickly draw the new comb for brood and stores.

The alternative to setting up a new hive or giving the bees away is to merge a swarm with another colony. Before doing this, check the bees are healthy and then find and remove the queen that led the swarm. Uniting two colonies is a great way to instantly strengthen a weak colony.

Prepare to harvest

By early summer, the build up of honey will be well underway with two or three supers on the hive. Keep up with the regular weekly inspections for a few more weeks, monitoring for signs of swarming. By the middle of summer the swarming season is coming to a close and you can begin to relax and enjoy the satisfaction of knowing that your bees are at full strength, busily capping the ripe honey in preparation for your harvest (see Time to harvest, p.68).

DID YOU KNOW
many garden flowers of the exotic or highly cultivated type produce little pollen or nectar and are of no use to bees at all.

As summer comes to an end, the main supply of nectar is on the decline, leaving the bees to forage for any late-flowering willow herb and bramble. This lull before the arrival of the ivy pollen and nectar in early autumn leaves the bees with little to gather so after your annual harvest, which will leave the bees in an agitated state, pacify them by placing some thick syrup or baker's fondant in the hive. This will allow the colony to take the food down to prepare for their winter stores. Each hive will require 12 lbs. sugar melted in water (see p.113 for syrup recipe), but this does not have to be offered in one go and can be topped off as required.

Heather honey

Toward the end of the summer, some beekeepers transport their bees some distance to take advantage of fields where, for a few weeks, heather carpets the landscape in a mass of purple flowers. Heather honey is very different to other types of honey—it is highly regarded for its delicious flavor, attractive deep amber color, and distinctive, gelatinous quality.

Autumn and winter

It is important to check and prepare your bees for the winter ahead. Honeybees do not hibernate, but spend the winter as a colony, unlike bumble bees and wasps where only the queen survives to hibernate. It is vital for their survival that the bees are left with adequate food stores and supplies after the honey harvest.

Check for signs of disease

Learn about and get to know the signs of the various conditions that can affect the health of your bees. The better their condition, the greater their chances are of surviving a long, wet winter and re-emerging as a strong, healthy colony the following spring.

Remembering to check your hive for the parasitic mite, varroa, is very important (see Hive maintenance, p.63). Manage any infestation with one of the recommended treatments (Apistan or Bayvarol strips) that need to be placed in the hive in early autumn for six weeks and then removed. Do not be tempted to leave your hive untreated as it will weaken your bees and could lead to the loss of your colony within a couple of years. If you think your bees are showing signs of disease and are unsure how to proceed, do get the advice of an experienced beekeeper who will be able to recommend what course of action to take.

Maintaining a healthy hive
It is every beekeeper's duty to maintain a healthy hive, not only for the well-being of their own bees but to safeguard neighboring colonies, too.

Winter supplies

The bees will need a good amount of food to see them through the winter and spring. On average, a healthy colony will require about 40 lbs. of sealed stores. The additional feed given after the honey harvest will have been taken down and used to fill the frames in the supers. The brood combs will also be used to pack in supplies.

In early autumn, the common ivy produces vast amounts of pollen and nectar and on a warm day the bees will make the most of this important source of food. Ivy honey, because of its high glucose content compared to the majority of nectars, is inclined to crystalize very quickly and for this reason it is hard for the bees to break it down for food during the spring, but if syrup is available while they are collecting from the ivy the stores will be a mix of the two and will provide food that is easier to feed on when they emerge after the winter.

When the temperature falls, the bees will form a tight cluster to maintain the warmth inside the hive. Going into winter with a strong stock of bees will help survival rates at this point. The bees will still be moving about the hive to feed on the stores and, on warmer days, they will take cleansing flights outside to defecate.

Breeding stops at this time of year, resuming in early spring. Removal of the queen excluder is advised so that the queen is free to move up into the super. If the excluder is left in position she risks being isolated and left to perish in the brood chamber when the cluster moves upwards. Replace the queen excluder in early spring having checked that the queen is in the brood chamber.

Other than pests, disease, and lack of food, the major thing for a bee to avoid is damp conditions. By keeping your hive well ventilated, this can be overcome. First, check the roof of your hive is waterproof and remove the bee escape from the crown board. A trick many use is to place a matchstick under each corner of the board to gently elevate it,

TEMPERATURE CONTROL
The bees will work to keep their hive at a constant temperature of about 93°F.

allowing moisture to evaporate from the hive. If you have chosen to have an open-mesh floor for your hive this should provide adequate ventilation. Do not worry that your bees will get cold if you do this—as long as they are strong in numbers they can survive long periods of chilly weather.

Take precautions against pest damage to hives and colonies. Put mouse guards in place over the larger entrances before any cold snap occurs that may drive mice into the hive for the winter. Woodpeckers cause damage for some beekeepers, drilling holes through hives in the hunt for food. Enclosing the hive in a wire cage or wrapping it in chicken wire should deter them.

As a safeguard, use a brick or something similar on the roof of the hive to act as a stabilizer in rough weather. If your hive is located away from your home, try and visit it fairly regularly to make sure it has not been knocked out of shape by a passing badger or deer.

During wintery weather and snowfalls check the hive entrance, which may have become blocked with drifting snow or even iced over altogether. When snow is on the ground the bees can be fooled by the bright glare of the sun outside and are lured into the cold air, only to be instantly chilled and unable to return to the hive. Placing a piece of board at an angle over the entrance to the hive will offer enough shade to prevent the bees from venturing out.

Ideally, you will already have your hive positioned with the entrance turned away from the prevailing wind, to avoid draughts and driving rain.

Knowing you have done all you can to protect the welfare of your bees, you can sit back and enjoy learning more about the honeybee and beekeeping. Prepare for the spring by sorting out your equipment, mending hives, and cleaning old frames in preparation for the coming year.

Resources

For additional information about beekeeping, and for lists of US suppliers, please visit these Web sites:

American Beekeeping Federation (ABF)
www.abfnet.org

Betterbee
www.betterbee.org

www.b2byellowpages.com/directory/b2b_equipment/bee-keeping-equipment.shtml